U0921406

中国海洋统计年鉴

CHINA MARINE STATISTICAL YEARBOOK 2010

国 家 海 洋 局

Edited by

State Oceanic Administration,

People's Republic of China

海洋出版社

China Ocean Press

图书在版编目（CIP）数据

中国海洋统计年鉴. 2010/国家海洋局编著. --北京：海洋出版社，2011.2

ISBN978-7-5027-7958-0

Ⅰ. ①中… Ⅱ. ①国… Ⅲ. ①海洋—统计资料—中国—2010—年鉴Ⅳ. ①P7-66

中国版本图书馆 CIP 数据核字（2011）第 016947 号

责任编辑：王　溪

责任印刷：刘志恒

海洋出版社 出版发行

http: //www.oceanpress.com.cn

北京市海淀区大慧寺路 8 号　　　　邮编：100081

国家海洋信心中心印刷厂印刷　　　新华书店发行所经销

2011 年 3 月第 1 版　　2011 年 3 月第 1 次印刷

开本：787mm×1092mm　1/16　印张：20.75

字数：495 千字　　印数：1~1500 册

定价：168.00 元

发行部：(010)62147016　　　邮购部：(010)68038093　　　总编室：(010)62114335

《中国海洋统计年鉴》编委会

柯王俊　　中国船舶工业集团公司

刘　悦　　中国船舶重工集团公司

李　军　　中国石油天然气集团公司

刘　岩　　中国石油化工集团公司

陈伟杰　　中国海洋石油总公司

胡红江　　中国盐业总公司

王华俊　　中国有色金属工业协会

来　鹤　　辽宁省海洋与渔业厅

张俊杰　　河北省国土资源厅

孙连友　　天津市海洋局

鲁小兵　　山东省海洋与渔业厅

唐庆宁　　江苏省海洋与渔业局

朱铁民　　上海市海洋局

刘向东　　浙江省海洋与渔业局

陈泽銮　　福建省海洋与渔业厅

文　斌　　广东省海洋与渔业局

田凤鸣　　广西壮族自治区国土资源厅

黄良胜　　海南省海洋与渔业厅

董艳涛　　大连市海洋与渔业局

黄聿颂　　青岛市海洋与渔业局

吴建义　　宁波市海洋与渔业局

周鲁闽　　厦门市海洋与渔业局

梁俊乾　　深圳市农业和渔业局

《中国海洋统计年鉴》编辑部

主　任：

何广顺　　国家海洋信息中心副主任

翁立新　　国家海洋局政策法规和规划司副司长

副主任：

王晓惠　　国家海洋信息中心海洋经济部主任

沈　君　　国家海洋局政策法规和规划司海洋经济处处长

成　员：

王占坤　李长如　杨　娜　刘　彬　林香红

郭　越　赵　锐　宋维玲　蔡大浩　董　伟

周洪军　朱　凌　徐丛春　赵　鹏　李宜良（国家海洋信息中心）

冀渺一　庞　冲（国家海洋局政策法规和规划司海洋经济处）

特邀编辑：

李艳丽　　教育部

秦浩源　　科技部

甘　红　　国土资源部

郑文英　　交通运输部

魏新平　　水利部

郭　睿　　农业部

毛玉如　　环境保护部

刘建杰　　国家林业局

赵　成　　国家旅游局

崔胜先　　中国科学院

陆覃星　　中国地震局

吴万克　　中国气象局

Editorial Board of
China Marine Statistical Yearbook

Zhang Liping	Chinese Academy of Sciences
Xu Tieju	China Earthquake Administration
Wang Bangzhong	China Meteorological Administration
Ke Wangjun	China State Shipbuilding Corporation
Liu Yue	China Shipbuilding Industry Corporation
Li Jun	China National Petroleum Corporation
Liu Yan	China Petrochemical Group Corporation
Chen Weijie	China National Offshore Oil Industry Corporation
Hu Hongjiang	China National Salt Industry Corporation
Wang Huajun	China Nonferrous Metals Industry Association
Lai He	Department of Ocean and Fisheries of Liaoning Province
Zhang Junjie	Hebei Province Department of Land and Resources
Sun Lianyou	Tianjin Ocean Administration
Lu Xiaobing	Department of Oceanic and Fishery of Shandong Province
Tang Qingning	Jiangsu Provincial Ocean and Fisheries Bureau
Zhu Tiemin	Shanghai Municipal Ocean Bureau
Liu Xiangdong	Zhejiang Provincial Ocean and Fisheries Bureau
Chen Zeluan	Department of Oceanic and Fishery of Fujian Province
Wen Bin	Guangdong Provincial Oceanic and Fishery Administration
Tian Fengming	Department of Land and Resources of Guangxi
Huang Liangsheng	Department of Marine and Fishery of Hainan Province
Dong Yantao	Dalian Ocean and Fishery Administration
Huang Yusong	Qingdao Ocean and Fishery Administration
Wu Jianyi	Ningbo Ocean and Fishery Bureau
Zhou Lumin	Oceans and Fisheries Bureau of Xiamen
Liang Junqian	Shenzhen Provincial Bureau of Agriculture and Fishery

Editorial Department of China Marine Statistical Yearbook

编 者 说 明

一、《中国海洋统计年鉴》2010年版是一部全面反映2009年中华人民共和国海洋经济发展和海洋管理服务情况的资料性年鉴，全书为中英文对照。

二、本年鉴的统计资料范围为人们在海洋和沿海地区开发、管理、利用海洋资源和空间，发展海洋经济的生产和活动以及沿海地区的社会经济概况。地域范围为沿海地区、沿海城市和沿海地带，其排列顺序按《沿海行政区域分类与代码》(HY/T 094-2006)的顺序排列。

三、本年鉴内容包括综合资料、海洋经济核算、主要海洋产业活动、主要海洋产业生产能力、涉海就业、海洋科学技术、海洋教育、海洋环境保护、海洋行政管理及公益服务、全国及沿海社会经济、部分世界海洋经济统计资料等十一部分。

四、本年鉴根据《海洋统计报表制度》（国统制[2009]42号）和《海洋生产总值核算制度》（国统制[2008]116号），资料主要来源于沿海省、自治区、直辖市统计局、海洋厅（局）以及20个有关涉海部、局、总公司。

五、本年鉴中除特殊说明外，所有价值量指标均为当年价，除注明年份外，其他均为2009年数据，年鉴中每部分后附有主要海洋统计指标解释，对主要海洋统计指标的含义、统计范围和统计方法做了简要说明。统计数据中的其他说明置于表的下面。有续表的资料，如有注释均置于最后一张表的下面。

六、本年鉴表格中符号使用说明："…"表示数据不足本表最小计算单位数；"空格"表示该项统计指标数据不详或无该项数据；"#"表示其中的主要项；其他符号如"*"或"①"等表示本表后面有注释。

七、本年鉴中由于数字精确度的原因，四舍五入后部分分项值之和与合计值有微小差异。

八、本年鉴资料国内部分未包括香港特别行政区、澳门特别行政区和台湾省数据。

九、《中国海洋统计年鉴》在编汇过程中，得到了各有关单位的大力支持，我们在此表示衷心的感谢。本年鉴中如有疏漏和不妥之处，敬请读者批评指正。

《中国海洋统计年鉴》编辑部

Introduction

I. *China Marine Statistical Yearbook (2010)* is a data almanac which reflects in an all-round way the development of marine economy, marine management and service in the People's Republic of China in 2009, and it is a Chinese-English bilingual edition.

II. The Yearbook's statistics cover the production and activities in the marine and coastal areas in relation to the development, management and utilization of marine resources and space, and the development of marine socioeconomy. The regions covered are the coastal regions, coastal cities and coastal zones with coastlines, which are arranged in order according to the *Coastal Administrative Areas Classification and Codes* (HY/T 094—2006).

III. The data in the yearbook consist of 11 sections, namely, integrated data, marine economic accounting, major marine industrial activities, production capacity of major marine industries, ocean-related employment, marine science and technology, marine education, marine environmental protection, marine administration and public-good service, national and coastal socioeconomy, part of the world's marine economic statistics data.

IV. The Yearbook is based on the *Marine Statistics Report System* (Guotongzhi [2009] No. 42) and the *Ocean Gross Product Accounting System* (Guotongzhi [2008] No. 116), its data mainly come from the statistical bureaus and the oceanic administrations of the coastal provinces, autonomous regions, and municipalities directly under the Central Government as well as the 20 ocean-related ministries, bureaus and general corporations concerned.

V. Unless otherwise specified in the Yearbook, all the value indicators are given at the current price. Each section is attached by explanatory notes to the major marine statistical indicators, giving a brief explanation for the meaning, statistical range and statistical methods of the major marine statistical indicators. Other notes to the statistical data are listed below the tables. For the data with continued tables, annotations, if any, are put below the last table.

VI. The usage of symbols in the tables: "···" indicates the statistics smaller than the minimum calculation unit in the table; "Blank" indicates that the data of the statistical index is unknown for the time being or that there shouldn't be any data; "#" indicates the major items of the table; Other symbols, such as "*" or "①", indicate "see footnotes below".

VII. For reasons of digital accuracy, there is small difference between the sum of values of some subterms after having been rounded off and the total values.

VIII. The domestic part of the Yearbook does not include that of Hong Kong Special Administrative Region, Macau Special Administrative Region and Taiwan Province.

IX. In the course of editing *China Marine Statistical Yearbook*, we enjoyed energetic support from the various departments concerned and we hereby extend our heartfelt thanks to them. Criticisms and comments are welcome from readers on any of the oversights and inappropriateness as the time for editing is too short.

Editorial Department of
China Marine Statistical Yearbook

目　次
CONTENTS

7　海洋教育

Marine Education

8 海洋环境保护
Marine Environmental Protection

10 全国及沿海社会经济
National and Coastal Socioeconomy

2009年我国海洋经济发展综述

2009年，各级海洋行政主管部门深入贯彻实践科学发展观，积极落实国家“保增长、调结构”的政策方针，努力克服国际金融危机等不利影响，全国海洋经济继续保持稳步增长态势。

一、全国海洋经济发展概况

2009年，全国海洋生产总值32 277.6亿元，比上年增长9.2%（除特别注明外，增长率均按可比价计算），海洋生产总值占国内生产总值的9.47%，占沿海地区生产总值的15.6%。全国涉海就业人员3 271万人，占沿海地区就业人员比重略有下降。

二、主要海洋产业发展情况

2009年，主要海洋产业实现增加值12 843.6亿元，比上年增长9.5%，占海洋生产总值的39.8%，滨海旅游业和海洋交通运输业仍占主导地位。

海洋第一产业 2009年，沿海地区海洋渔业生产稳步增长，全年海洋渔业实现增加值2 440.8亿元，比上年增长10.7%。海水产品产量达2 880.54万吨，比上年增长3.5%，其中，海水养殖产量1 536.46万吨，比上年增长6.4%；海洋捕捞产量1 344.08万吨，比上年增长0.24%；远洋捕捞产量81.86万吨，比上年减少10.2%。

海洋第二产业 2009年，海洋第二产业继续保持良好的发展态势。随着一批沿海和海上风电场建设项目相继运营投产，海洋电力业继续保持高速增长，全年实现增加值20.8亿元，比上年增长79.8%。海洋工程建筑业实现增加值672.3亿元，比上年增长71.4%。海洋船舶工业在《船舶工业调整与振兴规划》的指导下克服重重困难，继续保持快速发展态势，全年实现增加值986.5亿元，比上年增长32.6%，造船完工量达4 439.31万综合吨，比上年增长近90%。海洋化工业全年实现增加值465.3亿元，比上年增长24.8%。海洋矿业开采品种不断丰富，产量继续保持平稳增长，全年实现增加值41.6亿元，比上年增长33.7%。我国海水利用自主创新能力不断提升，大生活用水技术、海水利用装备制造等领域取得重大突破，全年实现增加值 7.8 亿元，比上年增长 2.5%。海洋盐业受下游两碱产业复苏的带动，全年海盐产量达 3 500.45 万吨，比上年增长 11.9%，实现增加值 43.6 亿元。海洋生物医药业全年实现增加值 52.1 亿元，比上年减少 8.3%。海洋油气业全年实现增加值 614.1 亿元，比上年减少 10.4％，全年海洋原油产量 3 698.19 万吨，比上年增长 8.1％，海洋天然气产量 859 173 万立方米，比上年增长 0.2％。

海洋第三产业　2009 年，海洋第三产业增速放缓，海洋交通运输业受金融危机影响较大，全年实现增加值 3 146.6 亿元，比上年减少 8％，沿海港口货物吞吐量达 487 371 万吨，比上年增长 8.6％，国际标准集装箱吞吐量 11 020 万标准箱，比上年减少 5.6％。滨海旅游业全年实现增加值 4 352.3 亿元，比上年增长 16.4％，全年滨海国际旅游收入达到 251.4 亿美元，比上年增加 32.4 亿美元，接待入境旅游者 4 558.54 万人次。

三、区域海洋经济发展情况

受国际金融危机冲击，2009 年成为新世纪以来我国海洋经济最为困难的一年。面对困难与挑战，从国家到沿海地区积极采取各项措施有效应对。辽宁沿海经济带、山东黄河三角洲高效生态经济区、江苏沿海地区、福建海峡西岸经济区、珠江三角洲地区纷纷上升为国家战略，为海洋经济的发展重新带来了机遇与活力。虽然 2009 年区域海洋经济增速明显减缓，但总量依然保持平稳增长的态势。环渤海经济区、长江三角洲经济区和珠江三角洲经济区海洋生产总值分别为 11 182.2 亿元、10 314.5 亿元和 6 661.0 亿元，分别占全国海洋生产总值 34.6%、32.0% 和 20.6%。

环渤海经济区海洋经济增速放缓，海洋生产总值比上年增长 6.6%（现价），占地区生产总值比重达 15.1%。海洋产业增加值为 6 292.6 亿元，海洋相关产业增加值为 4 889.8 亿元。环渤海三大海洋支柱产业中，除海洋交通运输业受金融危机冲击增加值出现下滑，海洋渔业和滨海旅游业依然保持增长态势，三大海洋产业增加值合计达到 3 718.82

亿元，占该地区主要海洋产业增加值的 76.9%。作为海洋第二产业较为发达的区域，环渤海地区海洋第二产业受金融危机的影响更为突出，其中由于国际油价大跌，海洋油气业降幅较为明显，其增加值与上年相比下降了 36.0%（现价）。海洋盐业和海洋化工业增加值也或多或少出现回落。海洋生物医药业、海洋电力业、海水利用业等海洋新兴产业显示了巨大的发展潜力，与上年相比增幅明显。

长江三角洲经济区增速继续减缓，2009年，海洋生产总值比去年增长9.1%（现价），比上年回落5个百分点，海洋生产总值占地区生产总值比重14.2%，与上年基本持平。长江三角洲海洋产业增加值为5 964.4亿元，海洋相关产业增加值为4 350.2亿元。在长三角四大支柱产业中，除海洋交通运输业受金融危机影响呈现负增长外，滨海旅游业、海洋船舶和海洋渔业继续保持增长态势，四大支柱产业占该地区主要海洋产业增加值的92.9%。此外，海洋电力业、海洋工程建筑业等海洋新兴产业增长迅速，增长速度分别达到109.5%和146.5%（现价）。

2009 年，珠江三角洲海洋经济区海洋经济继续快速增长，成为引领海洋经济快速发展的主力军。2009 年，珠江三角洲海洋生产总值达 6 661.0 亿元，比上年增长 5.1%，占地区生产总值比重达 16.9%。海洋产业增加值为 4 226.2 亿元，海洋相关产业增加值为 2 434.7 亿元。海洋产业中发展得较好的主要为滨海旅游业、海洋交通运输业、海洋化工业、海洋油气业和海洋渔业，其增加值之和占该地区主要海洋产业增加值 92.7%。

四、海洋科研教育管理服务情况

2009 年，海洋科研教育事业蓬勃发展，海洋科研机构共 186 个，从业人员 34 076 人，海洋科研项目 12 600 项，发表海洋科技论文 14 451 篇；出版海洋科技著作 248 种。开设海洋专业的高等学校共 314 个，专任教师 229 988 人，高等教育和中等职业教育海洋专业毕业生 67 214 人，招生 96 993 人，在校生 275 536 人，毕业班学生 87 579 人，海洋教育为全国海洋事业的发展培养了大批人才。

2009年，我国海洋环境保护各项工作稳步推进，沿海各地工业废水排放监管工作成效显著，11个沿海地区工业废水排放总量为1 388 343万吨，比上年减少5.5%，工业废水排放达标率为96.1%，比上年提高2.6个百分点；沿海各地日益重视海洋类型自然保护区的建设情况，共有保护区157个，面积29 460.77平方公里；我国近岸海域生态系统健康状况恶化的趋势依然存在，全国沿海对18个海洋生态监控区开展了监测，处于健康、亚健康和不健康状态的海洋生态系统分别占24%、52%和24%；全国污染海域面积增加，全海域未达到清洁海域水质标准的面积约为14.7万平方公里，比上年增长7.4%，轻度污染海域面积有所减少，但中度污染和严重污染海域面积均有所增加，海水水质污染面临严峻的形势；全年由于风暴潮灾害导致受灾人口872.12万人，比上年减少一半多，海水养殖受灾面积99.85千公顷，比上年增长了41%；全海域共发现赤潮68次，累计面积约14 100平方公里，与上年相比基本持平。

2009年，海洋行政管理工作开展有序，海洋开发利用秩序明显改善，海洋公益服务水平逐步提高。全年颁发海域使用权证书5 327本，确权海域面积17.84万公顷；全年共签发疏浚物海洋倾倒许可证104份，实施各项海洋执法检查共67 499次，发现违法行为1 817起，比上年减少36.6%；提供海洋数值预报服务共36 906万次，开展海洋调查项目4 051项，获得数据共654 143个，各项海洋监测获得数据量共9 280.9万个，全年接收存档卫星遥感数据量共计28 734.81 GB；室（馆）存海洋档案案卷113 621卷（册），磁介质档案9 122盘，全年共接待读者2 596人次。

Summary of National Marine Economic Development in 2009

In 2009, the competent marine administrative departments at all levels make an in-depth study of and put into practice the concept of scientific development, actively carry out the policy of "ensuring growth and adjusting structures", make great effort to conquer the effects of the international financial crisis, and give an impetus to the stable development of national marine economy.

I. Survey of the National Marine Economic Development

In 2009, the gross ocean product of the national marine economy reaches 3 227.76 billion Yuan, up 9.2% from the previous year (Unless otherwise specified, the growth rate is calculated in the comparable price.), accounting for 9.47% of the GDP, basically the same as that of previous year, and 15.6% of the coastal Gross Regional Product. The number of people employed by the ocean-related sectors reaches 32.71 million, whose proportion in the coastal employed personnel has slightly decreased from the previous year.

II. Development Situation of Major Marine Industries

In 2009, major marine industries effect an added value of 1 284.36 billion Yuan, up 9.5% from the previous year, accounting for 39.8% of the GOP, and the coastal tourism and marine communications and transportation industry still play a dominant role.

Primary marine industry In 2009, the marine fishery production of the coastal region increases steadily. The full-year marine fishery effects an added value of 244.08 billion Yuan, up 10.7% from the previous year; the output of seawater products amounts to 28.805 4 million tons, registering an increase rate of 3.5%, among which, the production of mariculture amounts to 15.364 6 million tons, up 6.4% from the previous year; the yield from marine fishing 13.440 8 million tons, up 0.24% from the previous year; and that from the deep-sea fishing 0.818 6 million tons, 10.2% down from the previous year.

Secondary marine industry In 2009, the secondary marine industry continues to keep a favourable posture of development. The marine electric power industry continues to keep a high speed increase. With a number of coastal and offshore wind electric field construction items put into production in succession, the coastal wind and tidal power generation effect an added value of 2.08 billion Yuan for the whole year and up 79.8% from the previous year. The marine engineering architecture effects a full-year added value of 67.23 billion Yuan, up 71.4% from the previous year. Under the guidance of the "*Shipbuilding Industry Adjustment and Promotion Plan*", the marine shipbuilding industry overcomes a series of difficulties, and maintains a rapid development trend. The full-year added value reaches 98.65 billion Yuan, increasing by 32.6%, and the completed quantity of ships built reaches 44.393 1 million tons, increasing by 90%. The marine chemical industry accomplishes an added value of 46.53 billion Yuan for the whole year, 24.8% up from the previous year. The exploitation variety of marine mining industry has become continuously enriched. Its production continues to keep a smooth increase, effecting a full-year added value of 4.16 billion Yuan, up 33.7% from the previous year. The self-pioneering capacity of seawater utilization has been enhanced continuously, and important breakthroughs have been made in the fields of large domestic water utilization technology and the manufacture of seawater utilization equipment, etc. and the sea water desalination and multipurpose utilization industry effects a full-year added value of 780 million Yuan, up 2.5% from the previous year. Spurred by the recovery of the downstream industry of soda ash and caustic soda, the marine salt industry effects a full-year added value of 4.36 billion Yuan and the production of sea salt amounts to 35.004 5 million tons, up 11.9% from the previous year. The marine biomedicine industry accomplishes an added value of 5.21 billion Yuan for the whole year, 8.3% down from the previous year. The offshore oil and gas industry effects a full-year added value of 61.41 billion Yuan, 10.4% down from the previous year. The annual crude oil output is 36.981 9 million tons, 8.1% up from the previous year, and the output of marine natural gas amounts to 8.591 73 billion cubic metres, 0.2% up from the previous year.

Tertiary marine industry In 2009, the growth rate of the tertiary marine industry slows down, and the effect of the international financial crisis on the marine communications and transportation industry is great, and an added value of 3 14.66 billion Yuan is accomplished for the whole year, 8% down for the previous year. The cargo handling capacity of coastal harbors amounts to 4 873.71 million tons, 8.6% up from the previous year and the handling capacity of international standardized containers 110.20 million standard cases, 5.6% down from the previous year. The coastal tourism effects a full-year added value of 435.23 billion Yuan, 16.4% up from the previous year, the full-year earnings from the coastal international tourism reaches $ 25.14 billion, $ 3.24 billion more than last year, and the number of inbound tourists received is 45.585 4 million person-times.

III. Situation of Regional Marine Economic Development

The year of 2009 might be the hardest year of the national marine economic development since 2000 due to the impact of the international financial crisis. However, facing all the difficulties and challenges, the State as well as coastal regions actively cope with them. The development of the economic zones, such as Liaoning Coastal Economic Zone, Shandong Yellow River Delta High-Effective Eco-Economic Zone, Jiangsu Coastal Zone, Fujian Economic Zone on the West Side of the Taiwan Straits, and Zhujing River Delta Economic Zone, have been upgraded to the national strategy, bringing new opportunities and vitality to the development of marine economy. Although the rate of regional marine economic growth in 2009 has obviously slowed down, the aggregate of it still presents a posture of stable growth. The gross ocean products of the Round-the-Bohai Economic Zone, Changjiang River Delta Economic Zone and Zhujiang

River Delta Economic Zone are 1 118.22 billion Yuan, 1 031.45 billion Yuan and 666.1 billion Yuan respectively, accounting for 34.6%, 32.0% and 20.6% of the national gross ocean product, respectively.

The growth rate of marine economy in the Round-the-Bohai zone has slowed down. The Gross Ocean Product increases by 6.6% (calculated at the current price), the growth rate was far below that of last year, and accounts for 15.1% of the Gross Regional Product. The added value of marine industries is 629.26 billion Yuan and that of ocean-related industries 488.98 billion Yuan. Among the three major marine pillar industries, marine fishery and coastal tourism have maintained a tendency to grow, except the marine communications and transportation industry which has been impacted by the finical crisis, and whose added value has dropped to some extent. The added value of three major marine industries amounts to 371.882 billion Yuan, accounting for 76.9% of the added value of major marine industries in the region, and the backbone role has been shown obviously. As a developed region of the secondary marine industries, the secondary industry in the Round-the-Bohai zone has been affected by the finical crisis more seriously. The rate of decrease of the offshore oil and gas industry is obvious due to the sharp decline of international oil price, and the added value has decreased by 36.0% (calculated at the current price). The added value of the marine industries such as marine salt industry and marine chemical industry have shown more or less a dropping trend. The new industries such as marine biomedicine, marine electric power and seawater initialization, etc., have shown huge potential, for development and their growth rates are obvious as compared with lost year.

The growth rate of the marine economy in the Changjiang River Delta Economic Zone continues to slow down. In 2009, the Gross Ocean Product increases by 9.1% as against that last year (at the current price), falling for nearly 5 percentage points and its proportion in the Gross Regional Product is 14.2%, basically the same as that last year. The added value of marine industries is 596.44 billion Yuan and that of ocean-related industries 435.02 billion Yuan. Among the four pillar industries, except the marine communications and transport industry which has shown negative growth due to the impact of the financial crisis, coastal tourism, marine shipbuilding, and marine fishery continue to grow, accounting for 92.9% of the added value of major marine industries in the region. And the growth rates of new industries such as marine electric power and marine engineering architecture, etc., are 109.5% and 146.5% (at the current price), respectively.

In 2009, the marine economy in the Zhujiang River Delta Economic Zone continues to grow rapidly and becomes the main force leading the rapid development of marine economy. In 2009, the gross ocean product of the Zhujiang River Delta Economic Zone reaches 666.10 billion Yuan, up 5.1% from the previous year, occupying a proportion of 16.9% in the Gross Regional Product. The added value of marine industries is 422.62 billion Yuan and that of the ocean-related industries 243.47 billion Yuan. The well-developed marine industries are coastal tourism, marine communications and transportation, marine chemical industry, offshore oil and gas industry, and marine fishery, and the added value of these industries accounts for 92.7% of the added value of major marine industries in the region.

IV. Marine Scientific Research Education and Public Service

In 2009, marine scientific research and education develop vigorously. The number of marine scientific research institutions is 186, and they employ 34 076 people. The number of marine scientific research projects reaches 12 600, and 14 451 marine scientific and technological papers and 248 kinds of marine scientific and technological works have been published. There are 314 ordinary marine colleges and universities that set up marine specialties, employing 229 988 full-time teachers. In the marine specialties of ordinary higher education and secondary vocational education there are 67 214 graduates, 96 993 at school, 275 536 new students and 87 579 in the graduating classes. Marine education has trained a large number of talents for the development of national marine undertakings.

In 2009, China's marine environmental protection develops steadily. The coastal industrial wastewater discharge monitoring in the coastal areas has got remarkable results, the total discharge of industrial waste water in the 11 coastal regions reaches 13.88343 billion tons, 5.5% down from the previous year; and the up-to-standard discharge rate of industrial waste water is 96.1%, 2.6 percentage points more than that last year; All coastal regions have increasingly paid attention to the construction of nature reserves of marine type and there are 157 protected areas with a total area of 29 460.77 km^2; The tendancy of the health condition of ecosystems in the Chinas near shore sea areas to deteriorate remains as usual. The nation's coastal regions have monitored 18 marine ecological monitoring areas, the proportion of healthy, sub-healthy and unhealthy state of the marine ecosystem is 24%, 52% and 24% respectively. The polluted sea area in the country has increased, the area that is not up to the standard of clean sea area water quality in the whole sea area is about 147 000 km^2, 7.4% up from the previous year, and the lightly polluted sea area has reduced, but the moderately-polluted and heavily-polluted sea areas have increased, and the seawater quality pollution is facing a severe situation; The storm surge disaster-stricken population reaches 8 721 200 people, over 50% down from the previous year, the affected area of mariculture is 99.85 thousand hectares, increasing by 41% over the previous year; a total of 68 times of red tide are found in the whole sea area, and the cumulative disaster area is about 14 100 km^2, basically the same as that in the previous year.

In 2009, marine administration is carried out in an orderly way, the order of ocean development and use has been obviously improved, and the level of marine public service is further raised. A total of 5 327 certificates for the sea area use are issued for the whole year, the sea area with the ownership of patent rights having reeached 178 400 hm^2; A total of 104 permits for oceanic dumping of dredged material are signed and issued, and various marine inspections for a enforcement are carried out on 67 499 occasions, in which 1 817 cases of unlawful practice are discovered, 36.6% down from the previous year; Marine numerical forecast service is provided on 369.06 million occasions, 4 051 marine survey items are carried out and 654 143 data are acquired, the data quantity obtained from various items of marine monitoring is 92.809 million and the satellite remote-sensing data received and placed on file a total 28 734.81 GB; The number of marine files kept in the divisions (Marine Archives) reaches 113 621 volumes (copies), with 9 122 disks of magnetic-medium file. For the whole year, the number of readers received amounts to 2 596 person-times.

图1 全国海洋生产总值及三次产业构成

China's Gross Ocean Product and Three Industries Composition

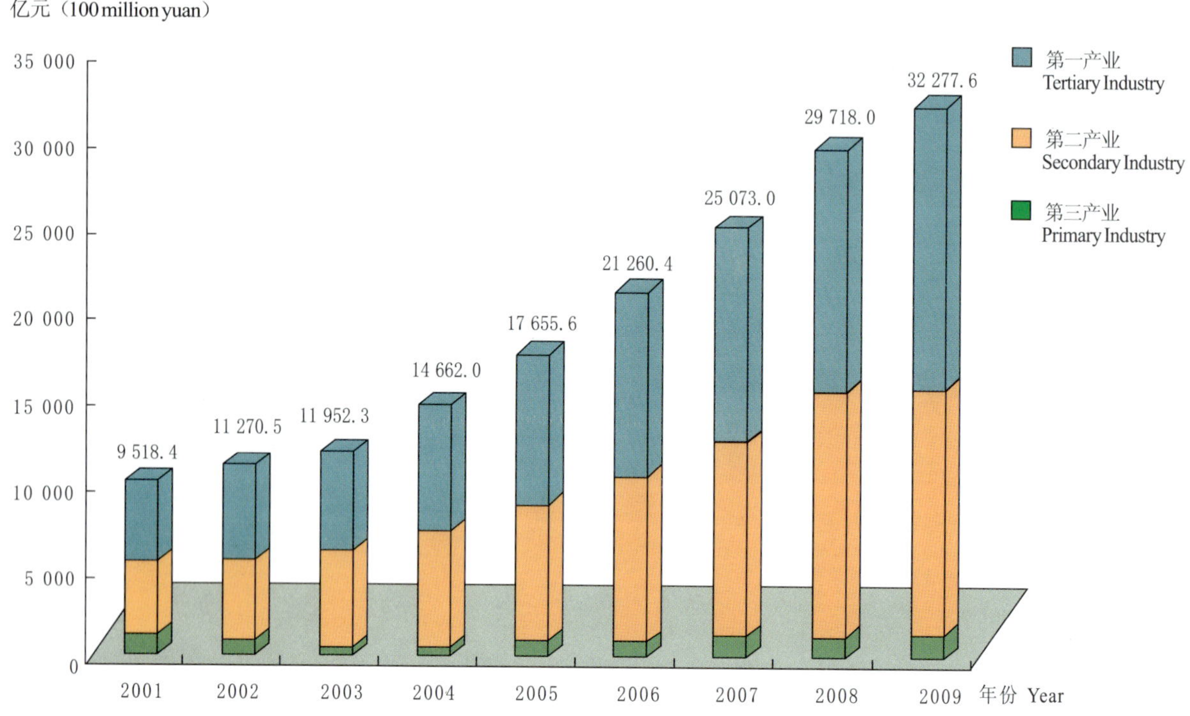

图2 2009年全国主要海洋产业增加值构成

Composition of Added Values of the Major Marine Industries in China in 2009

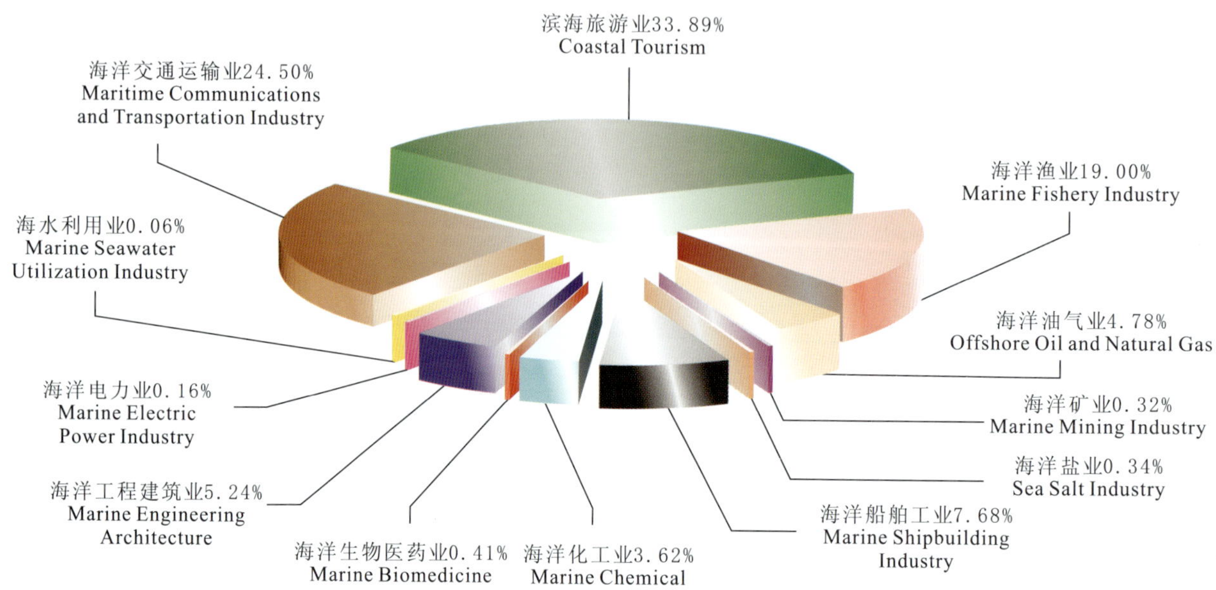

图3　2009年沿海地区海洋生产总值
Gross Ocean Product by Coastal Regions in 2009

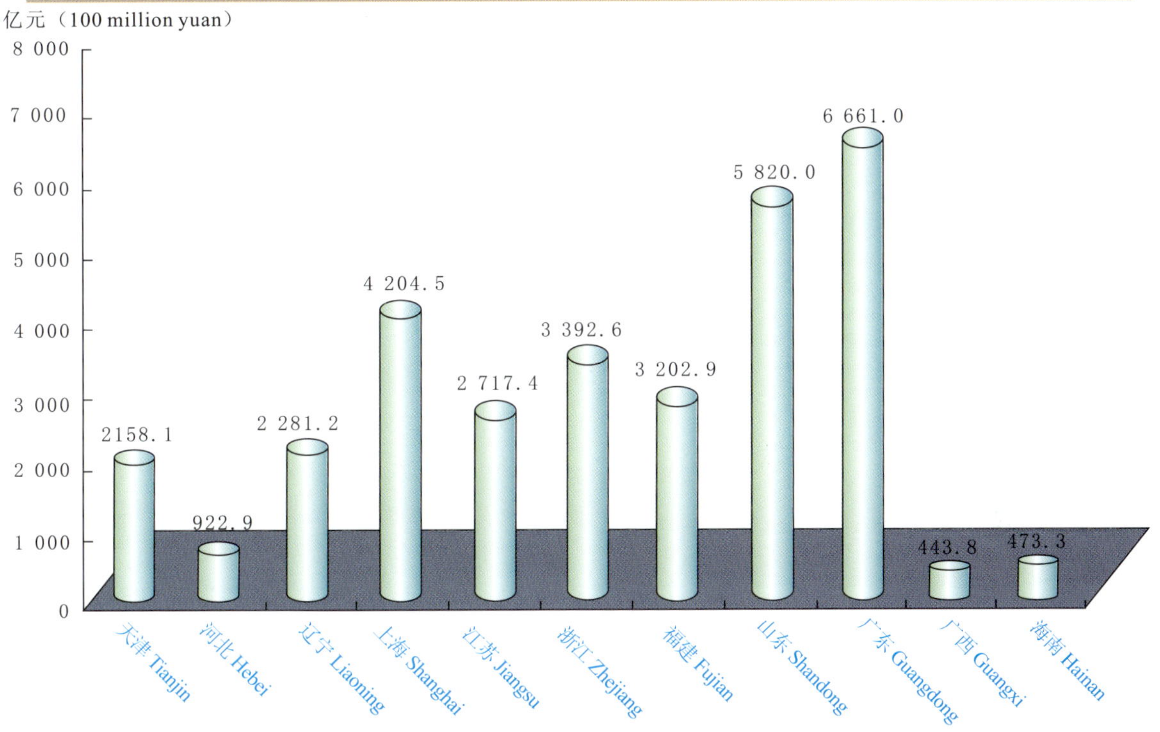

图4　全国海洋捕捞养殖产量
National Marine Catches and Mariculture Production

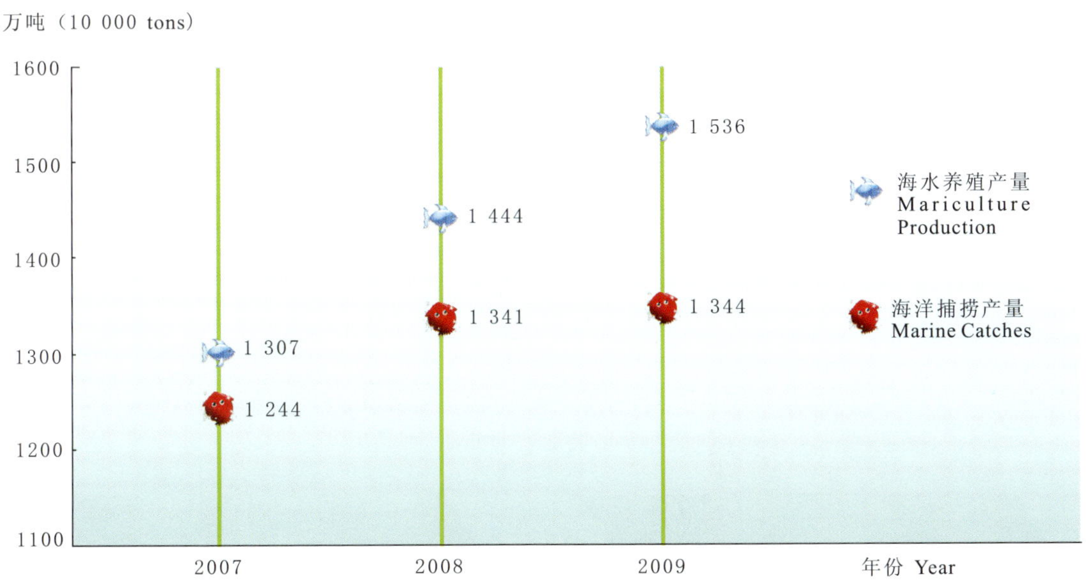

图5 2009年沿海地区海洋捕捞养殖产量

Marine Catches and Mariculture Production by Coastal Regions in 2009

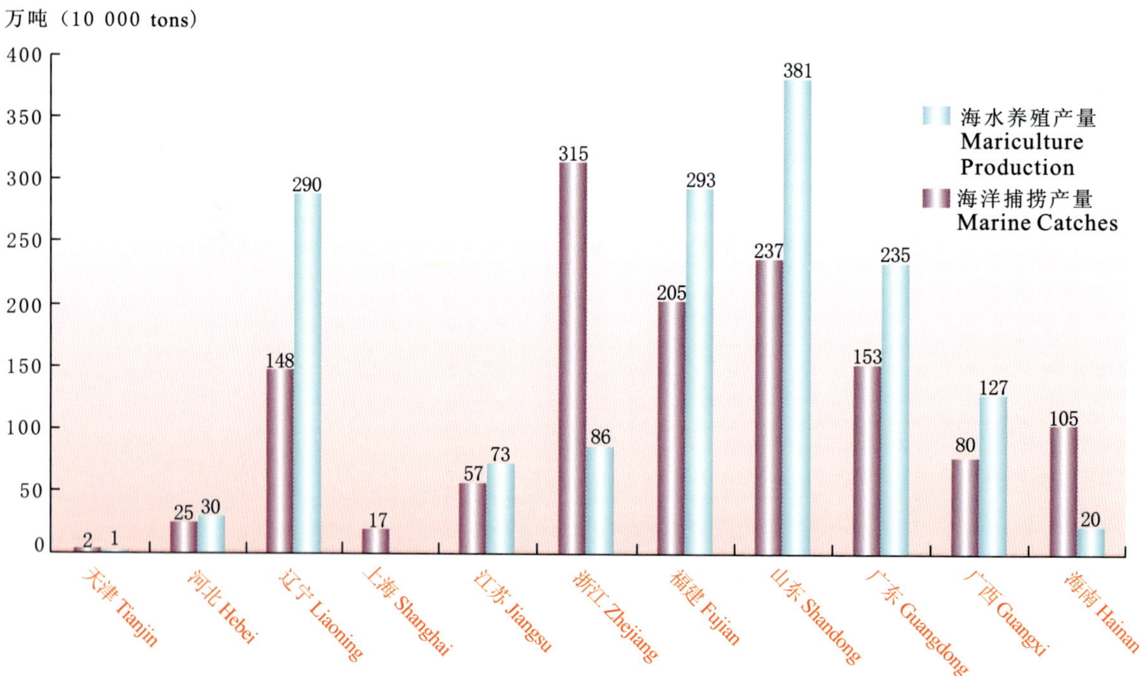

图6 全国海洋石油和天然气产量

National Output of Offshore Oil and Natural Gas

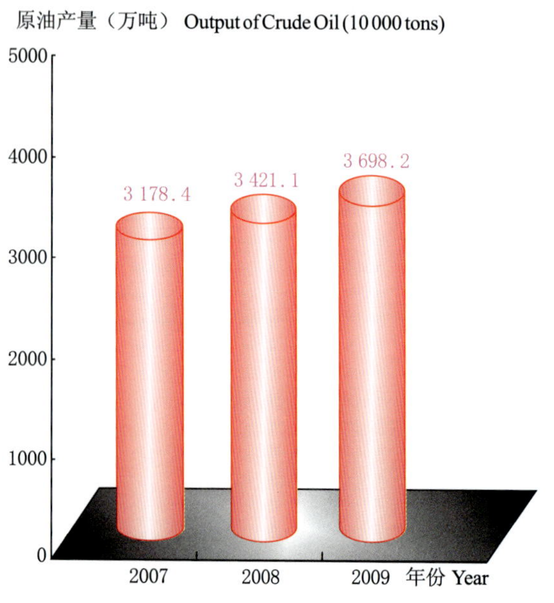

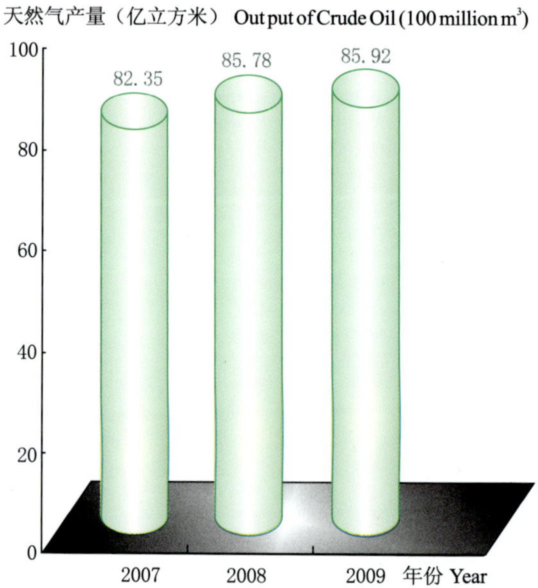

图7 海洋原油产量占全国原油产量比重

Proportion of Offshore Crude Oil Production in the National Total

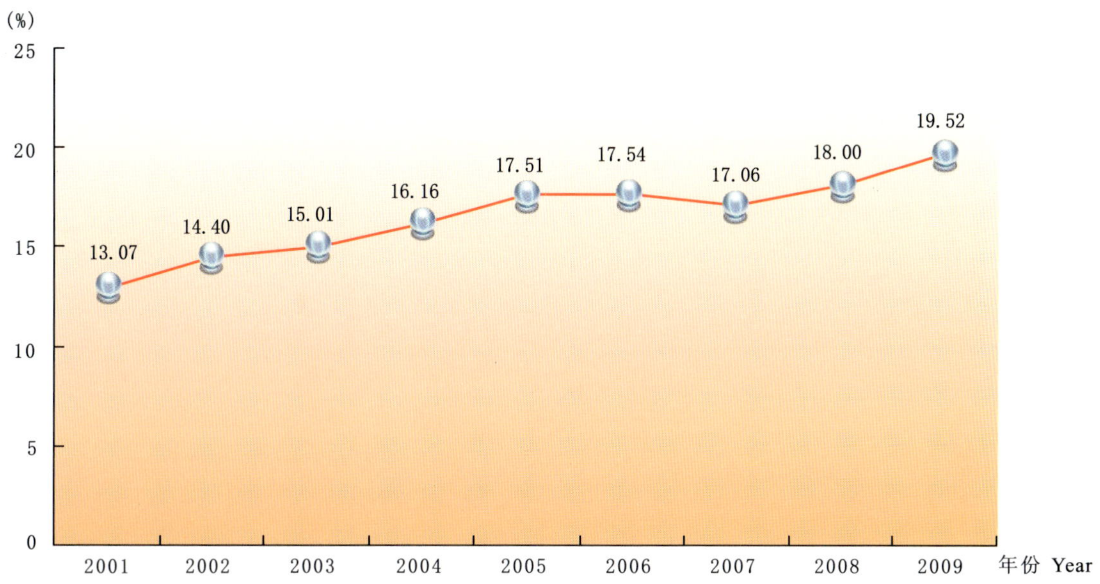

图8 2009年海洋矿业产量

Production of Marine Mining Industry in 2009

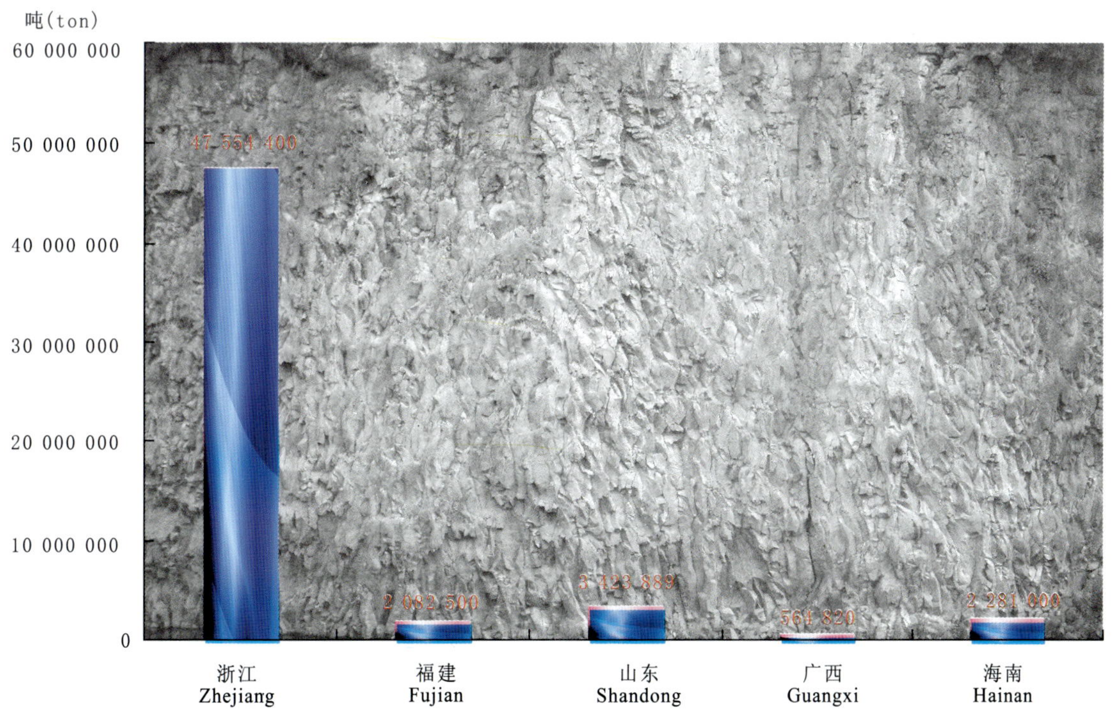

图9　2009年沿海地区海盐产量
Sea Salt Production by Coastal Regions in 2009

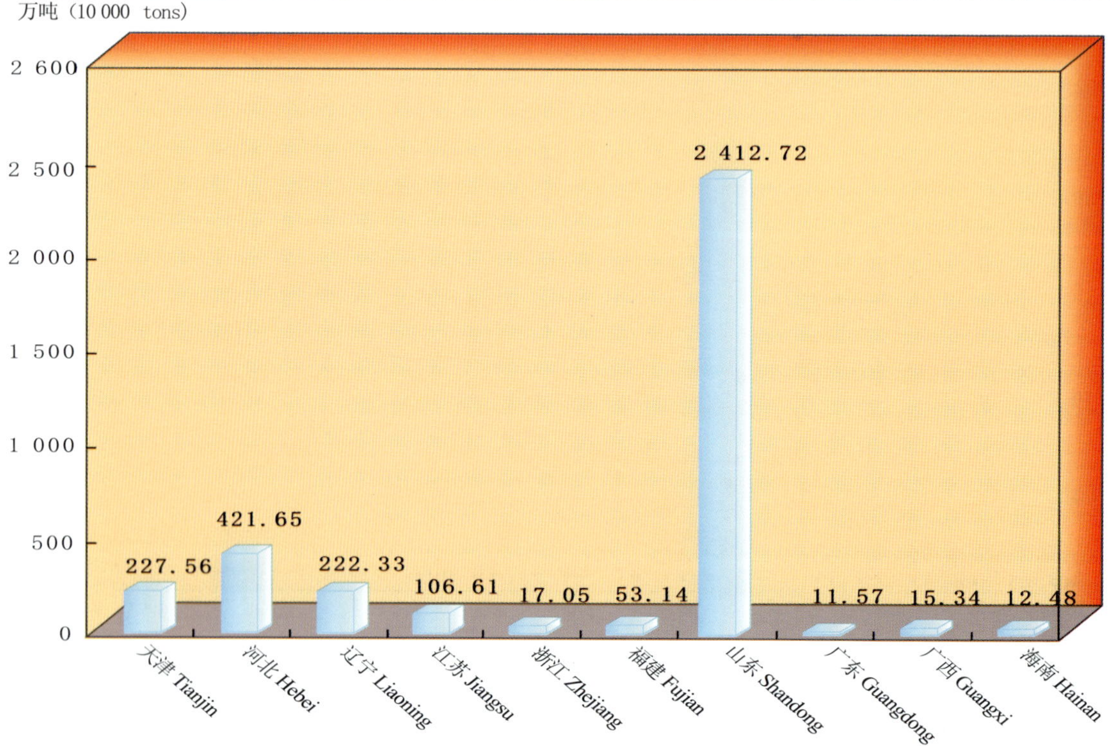

图10　2009年沿海地区造船完工量
Completed Number of Ships Built by Coastal Regions in 2009

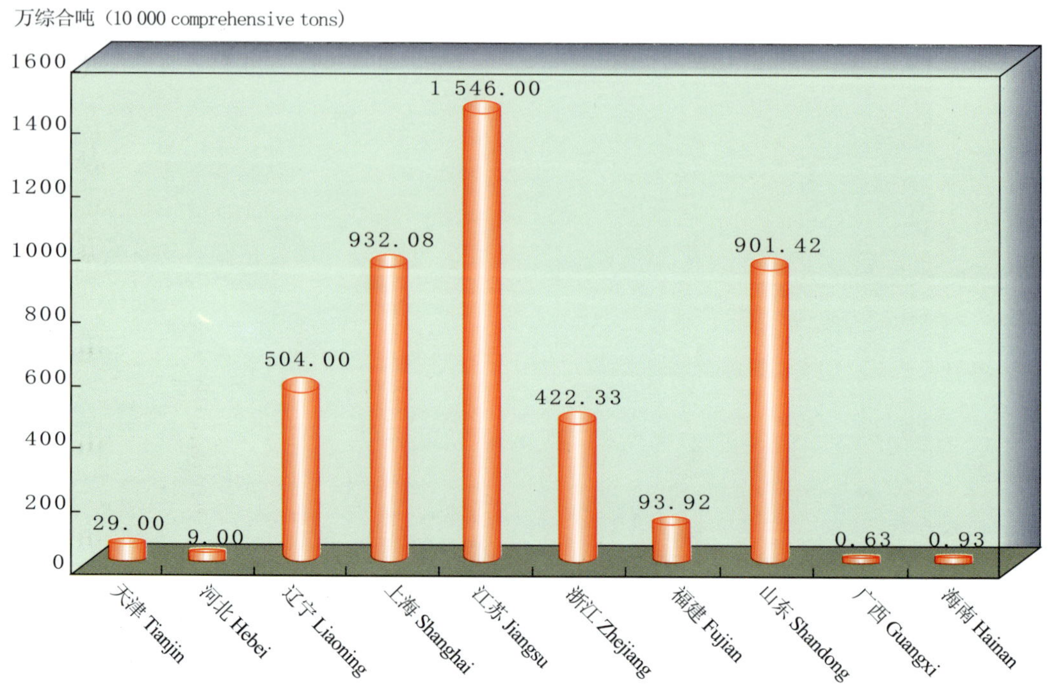

图11　2009年沿海地区海洋货物周转量
Goods Turnover Volume by Coastal Regions in 2009

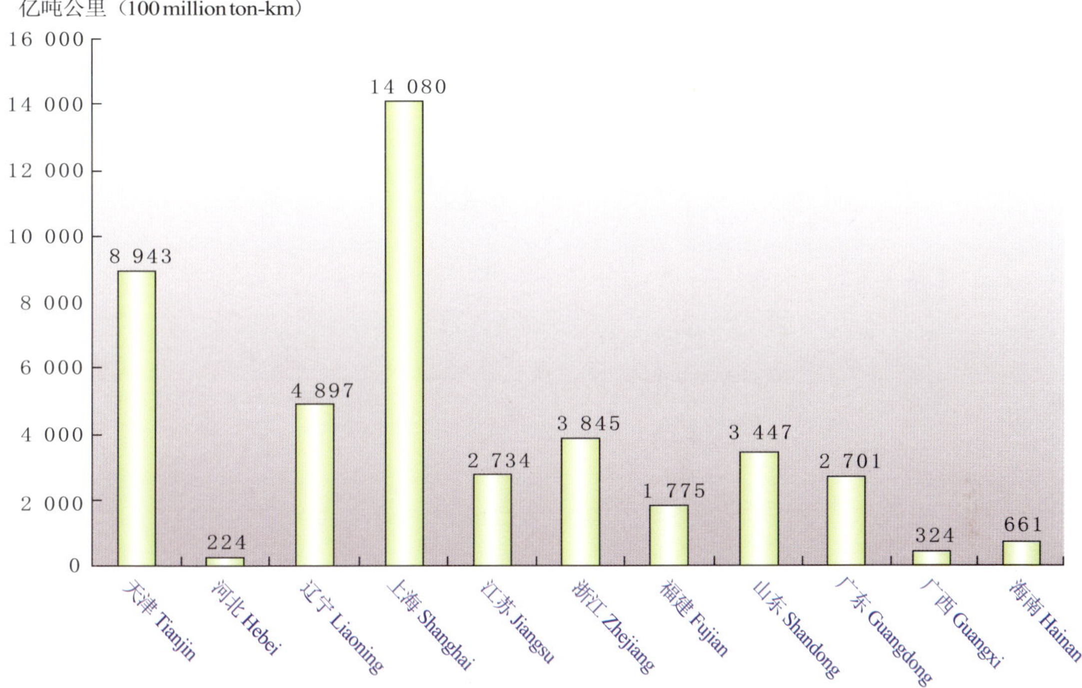

图12　2009年沿海港口国际标准集装箱吞吐量
International Standardized Containers Handled Coastal Seaports in 2009

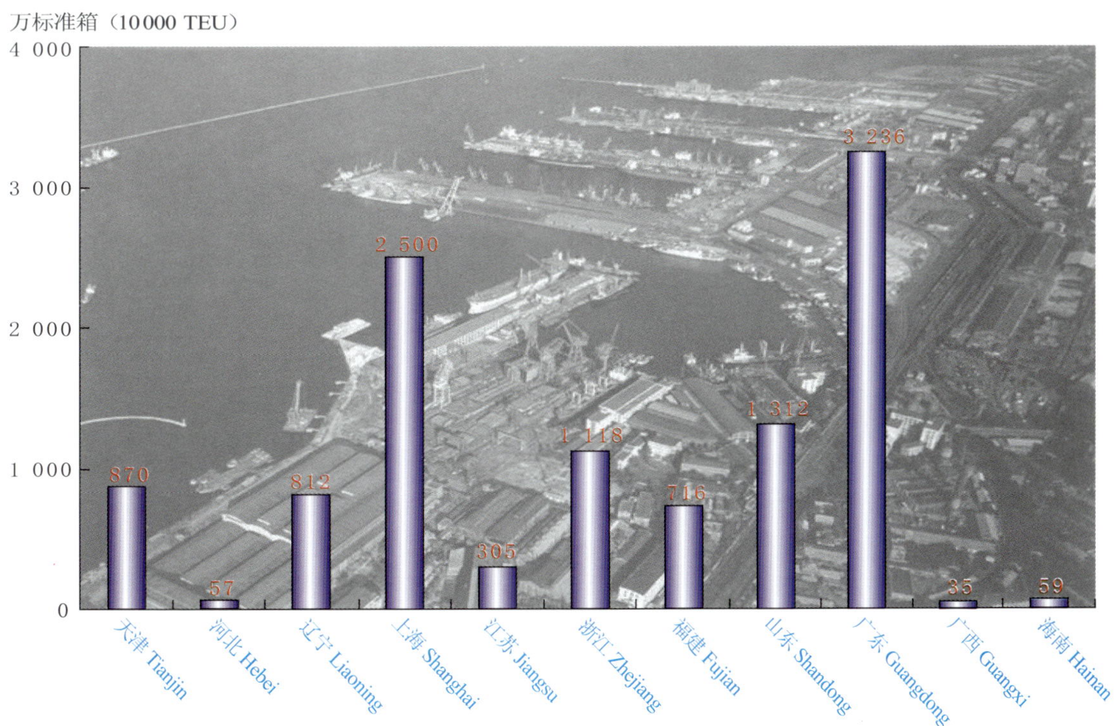

图13 沿海城市国际旅游收入
Earnings from International Tourism by Coastal Cities

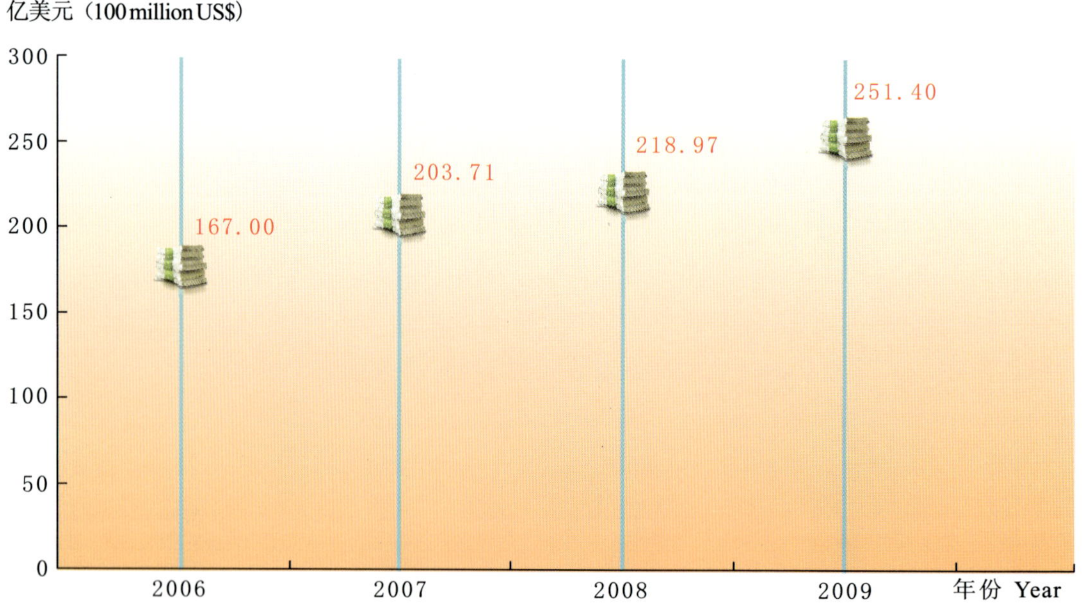

图14 沿海城市接待入境旅游者人数
Number of Inbound Tourists Received by Coastal Cities

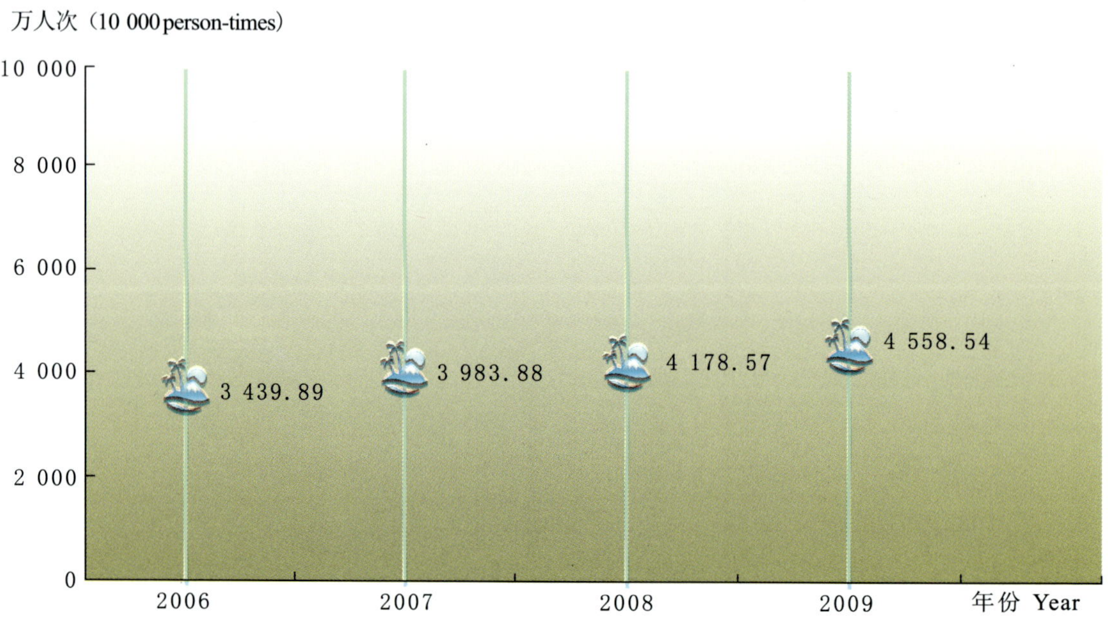

图15 2009年沿海地区海洋生产总值

Gross Ocean Product by Coastal Regions in 2009

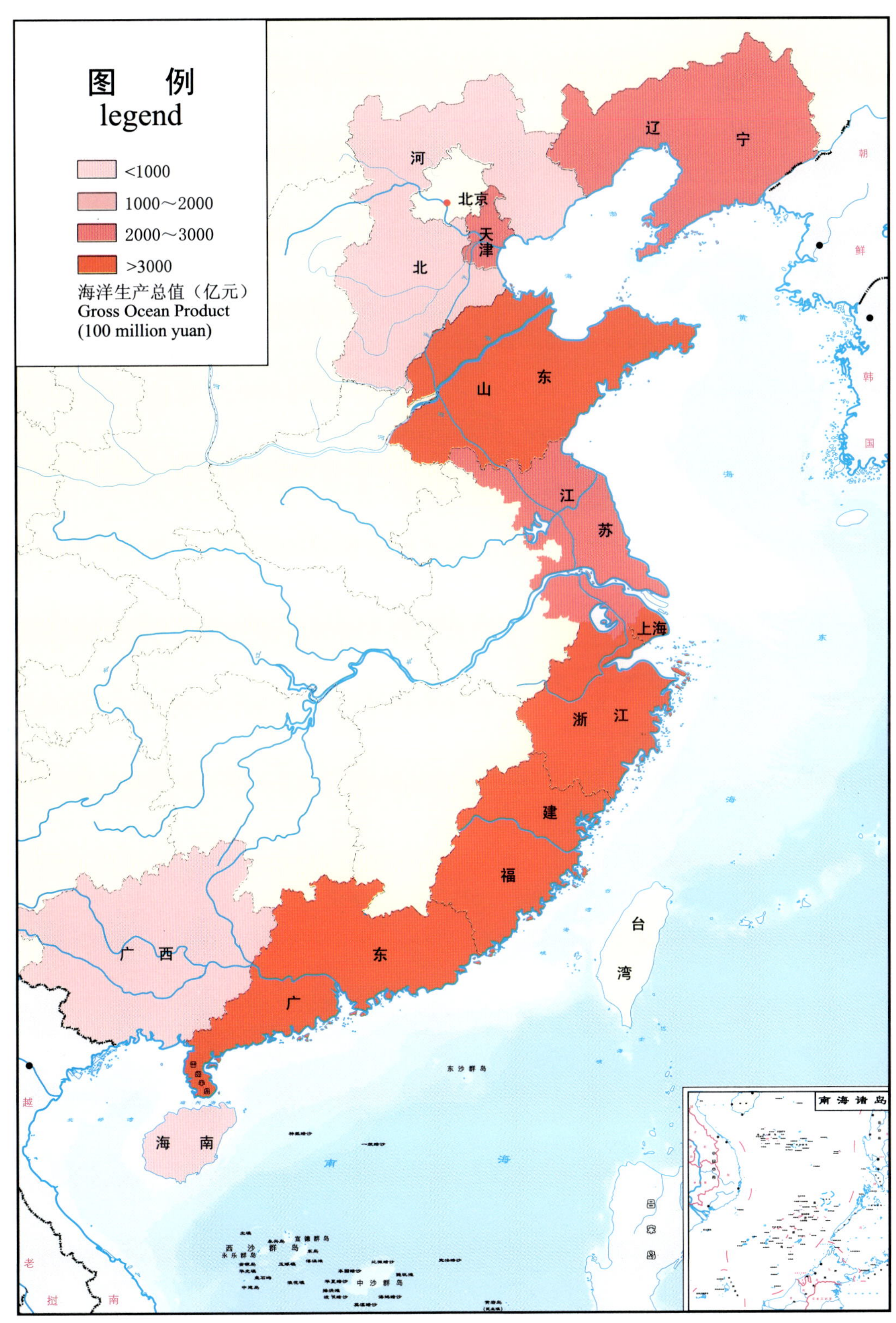

图16 2009年沿海地区海洋经济贡献

Marine Economic Contributions by Coastal Regions in 2009

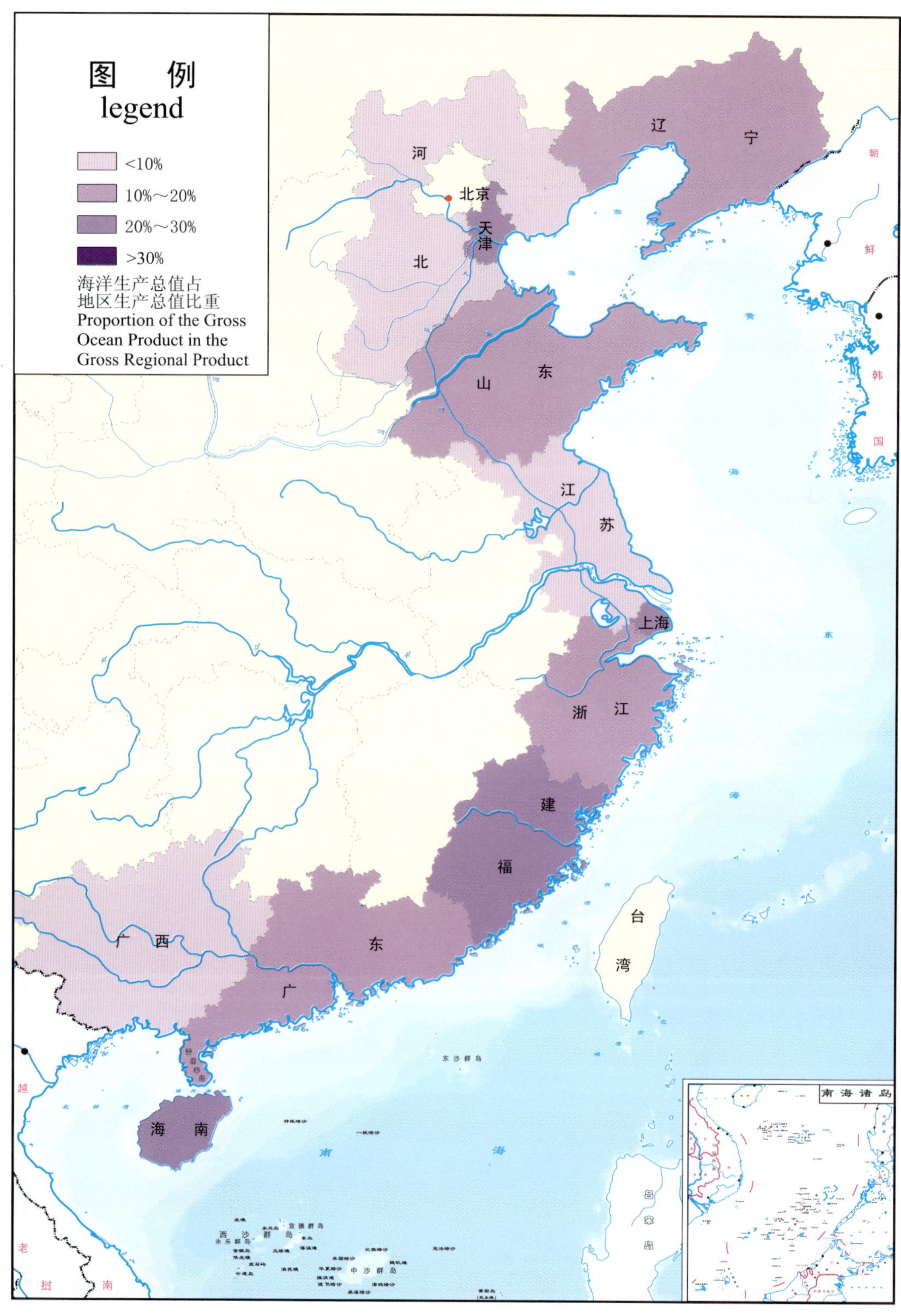

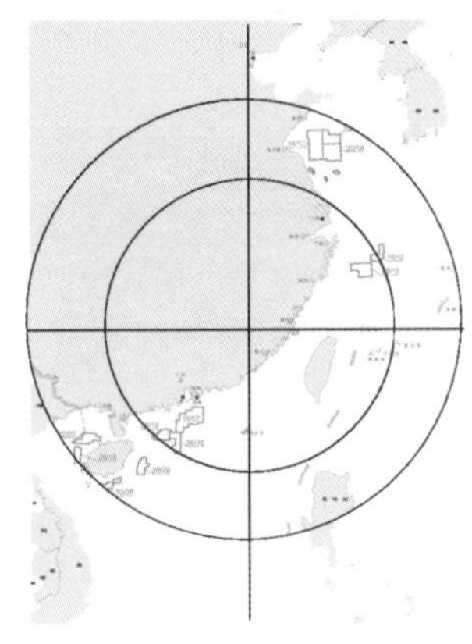

1

综 合 资 料

Integrated Data

1-1 沿海地区行政区划
Administrative Division of Coastal Regions

单位：个 (number)

沿海地区 Coastal Region	沿海城市 Coastal City	沿海地带 Coastal County (District)			
		合 计 Total	县 County	县级市 County-level city	区 District
合 计 Total	**53**	**237**	**64**	**56**	**117**
天 津 Tianjin	1	1			1
河 北 Hebei	3	11	6	1	4
辽 宁 Liaoning	6	22	4	7	11
上 海 Shanghai	1	5	1		4
江 苏 Jiangsu	3	15	8	4	3
浙 江 Zhejiang	7	35	11	10	14
福 建 Fujian	6	34	11	8	15
山 东 Shandong	7	37	6	14	17
广 东 Guangdong	14	56	11	6	39
广 西 Guangxi	3	8	1	1	6
海 南 Hainan	2	13	5	5	3

注：沿海地带中未包括广东省的东莞、中山和海南省的三亚。

Note: Dongguan, Zhongshan of Guangdong Province and Sanya of Hainan Province are not included in the Coastal County.

1-2 沿海行政区划一览表
Table of Administrative Division of Coastal Regions

沿海地区 Coastal Region	地区代码 Zip Code	沿海城市 Coastal City	地区代码 Zip Code	沿海地带 Coastal County (District)	地区代码 Zip Code
天 津 Tianjin	120000			滨海新区Binhai Xinqu	120116
河 北 Hebei	130000	唐山 Tangshan	130200	丰南区Fengnan Qu	130207
				滦南县Luannan Xian	130224
				乐亭县Leting Xian	130225
				唐海县Tanghai Xian	130230
		秦皇岛 Qinhuangdao	130300	海港区Haigang Qu	130302
				山海关区Shanhaiguan Qu	130303
				北戴河区Beidaihe Qu	130304
				昌黎县Changli Xian	130322
				抚宁县Funing Xian	130323
		沧州 Cangzhou	130900	海兴县Haixing Xian	130924
				黄骅市Huanghua Shi	130983
辽 宁 Liaoning	210000	大连 Dalian	210200	中山区Zhongshan Qu	210202
				西岗区Xigang Qu	210203
				沙河口区Shahekou Qu	210204
				甘井子区Ganjingzi Qu	210211
				旅顺口区Lüshunkou Qu	210212
				金州区Jinzhou Qu	210213
				长海县Changhai Xian	210224
				瓦房店市Wafangdian Shi	210281
				普兰店市Pulandian Shi	210282
				庄河市Zhuanghe Shi	210283
		丹东 Dandong	210600	东港市Donggang Shi	210681
		锦州 Jinzhou	210700	凌海市Linghai Shi	210781
		营口 Yingkou	210800	西市区Xishi Qu	210803
				鲅鱼圈区Bayuquan Qu	210804
				老边区Laobian Qu	210811
				盖州市Gaizhou Shi	210881
		盘锦 Panjin	211100	大洼县Dawa Xian	211121
				盘山县Panshan Xian	211122

1-2 续表1 continued

沿海地区 Coastal Region	地区代码 Zip Code	沿海城市 Coastal City	地区代码 Zip Code	沿海地带 Coastal County (District)	地区代码 Zip Code
		葫芦岛 Huludao	211400	连山区Lianshan Qu	211402
				龙港区Longgang Qu	211403
				绥中县Suizhong Xian	211421
				兴城市Xingcheng Shi	211481
上海 Shanghai	310000			宝山区Baoshan Qu	310113
				浦东新区Pudong Xinqu	310115
				金山区Jinshan Qu	310116
				奉贤区Fengxian Qu	310120
				崇明县Chongming Xian	310230
江苏 Jiangsu	320000	南通 Nantong	320600	通州区Tongzhou Qu	320612
				海安县Hai'an Xian	320621
				如东县Rudong Xian	320623
				启东市Qidong Shi	320681
				海门市Haimen Shi	320684
		连云港 Lianyungang	320700	连云区Lianyun Qu	320703
				新浦区Xinpu Qu	320705
				赣榆县Ganyu Xian	320721
				灌云县Guanyun Xian	320723
				灌南县Guannan Xian	320724
		盐城 Yancheng	320900	响水县Xiangshui Xian	320921
				滨海县Binhai Xian	320922
				射阳县Sheyang Xian	320924
				东台市Dongtai Shi	320981
				大丰市Dafeng Shi	320982
浙江 Zhejiang	330000	杭州 Hangzhou	330100	滨江区Binjiang Qu	330108
				萧山区Xiaoshan Qu	330109
		宁波 Ningbo	330200	海曙区Haishu Qu	330203
				江东区Jiangdong Qu	330204
				江北区Jiangbei Qu	330205
				北仑区Beilun Qu	330206
				镇海区Zhenhai Qu	330211
				鄞州区Yinzhou Qu	330212
				象山县Xiangshan Xian	330225
				宁海县Ninghai Xian	330226
				余姚市Yuyao Shi	330281
				慈溪市Cixi Shi	330282
				奉化市Fenghua Shi	330283

1-2 续表2 continued

沿海地区 Coastal Region	地区代码 Zip Code	沿海城市 Coastal City	地区代码 Zip Code	沿海地带 Coastal County (District)	地区代码 Zip Code
		温州 Wenzhou	330300	龙湾区Longwan Qu	330303
				瓯海区Ouhai Qu	330304
				洞头县Dongtou Xian	330322
				平阳县Pingyang Xian	330326
				苍南县Cangnan Xian	330327
				瑞安市Rui'an Shi	330381
				乐清市Yueqing Shi	330382
		嘉兴 Jiaxing	330400	海盐县Haiyan Xian	330424
				海宁市Haining Shi	330481
				平湖市Pinghu Shi	330482
		绍兴 Shaoxing	330600	绍兴县Shaoxing Xian	330621
				上虞市Shangyu Shi	330682
		舟山 Zhoushan	330900	定海区Dinghai Qu	330902
				普陀区Putuo Qu	330903
				岱山县Daishan Xian	330921
				嵊泗县Shengsi Xian	330922
		台州 Taizhou	331000	椒江区Jiaojiang Qu	331002
				路桥区Luqiao Qu	331004
				玉环县Yuhuan Xian	331021
				三门县Sanmen Xian	331022
				温岭市Wenling Shi	331081
				临海市Linhai Shi	331082
福 建 Fujian	350000	福州 Fuzhou	350100	马尾区Mawei Qu	350105
				连江县Lianjiang Xian	350122
				罗源县Luoyuan Xian	350123
				平潭县Pingtan Xian	350128
				福清市Fuqing Shi	350181
				长乐市Changle Shi	350182
		厦门 Xiamen	350200	思明区Siming Qu	350203
				海沧区Haicang Qu	350205

1-2 续表3 continued

沿海地区 Coastal Region	地区代码 Zip Code	沿海城市 Coastal City	地区代码 Zip Code	沿海地带 Coastal County (District)	地区代码 Zip Code
				湖里区Huli Qu	350206
				集美区Jimei Qu	350211
				同安区Tong'an Qu	350212
				翔安区Xiang'an Qu	350213
		莆田 Putian	350300	城厢区Chengxiang Qu	350302
				涵江区Hanjiang Qu	350303
				荔城区Licheng Qu	350304
				秀屿区Xiuyu Qu	350305
				仙游县Xianyou Xian	350322
		泉州 Quanzhou	350500	丰泽区Fengze Qu	350503
				洛江区Luojiang Qu	350504
				泉港区Quangang Qu	350505
				惠安县Hui'an Xian	350521
				金门县Jinmen Xian	350527
				石狮市Shishi Shi	350581
				晋江市Jinjiang Shi	350582
				南安市Nan'an Shi	350583
		漳州 Zhangzhou	350600	云霄县Yunxiao Xian	350622
				漳浦县Zhangpu Xian	350623
				诏安县Zhao'an Xian	350624
				东山县Dongshan Xian	350626
				龙海市Longhai Shi	350681
		宁德 Ningde	350900	蕉城区Jiaocheng Qu	350902
				霞浦县Xiapu Xian	350921
				福安市Fu'an Shi	350981
				福鼎市Fuding Shi	350982
山 东 Shandong	370000	青岛 Qingdao	370200	市南区Shinan Qu	370202
				市北区Shibei Qu	370203
				四方区Sifang Qu	370205
				黄岛区Huangdao Qu	370211
				崂山区Laoshan Qu	370212
				李沧区Licang Qu	370213
				城阳区Chengyang Qu	370214
				胶州市Jiaozhou Shi	370281
				即墨市Jimo Shi	370282
				胶南市Jiaonan Shi	370284

1-2 续表4 continued

沿海地区 Coastal Region	地区代码 Zip Code	沿海城市 Coastal City	地区代码 Zip Code	沿海地带 Coastal County (District)	地区代码 Zip Code
		东营 Dongying	370500	东营区Dongying Qu	370502
				河口区Hekou Qu	370503
				垦利县Kenli Xian	370521
				利津县Lijin Xian	370522
				广饶县Guangrao Xian	370523
		烟台 Yantai	370600	芝罘区Zhifu Qu	370602
				福山区Fushan Qu	370611
				牟平区Muping Qu	370612
				莱山区Laishan Qu	370613
				长岛县Changdao Xian	370634
				龙口市Longkou Shi	370681
				莱阳市Laiyang Shi	370682
				莱州市Laizhou Shi	370683
				蓬莱市Penglai Shi	370684
				招远市Zhaoyuan Shi	370685
				海阳市Haiyang Shi	370687
		潍坊 Weifang	370700	寒亭区Hanting Qu	370703
				寿光市Shouguang Shi	370783
				昌邑市Changyi Shi	370786
		威海 Weihai	371000	环翠区Huancui Qu	371002
				文登市Wendeng Shi	371081
				荣成市Rongcheng Shi	371082
				乳山市Rushan Shi	371083
		日照 Rizhao	371100	东港区Donggang Qu	371102
				岚山区Lanshan Qu	371103
		滨州 Binzhou	371600	无棣县Wudi Xian	371623
				沾化县Zhanhua Xian	371624
广 东 Guangdong	440000	广州 Guangzhou	440100	荔湾区Liwan Qu	440103
				越秀区Yuexiu Qu	440104
				海珠区Haizhu Qu	440105
				天河区Tianhe Qu	440106
				白云区Baiyun Qu	440111
				黄埔区Huangpu Qu	440112
				番禺区Panyu Qu	440113
				南沙区Nansha Qu	440115
				萝岗区Luogang Qu	440116

1-2 续表5 continued

沿海地区 Coastal Region	地区代码 Zip Code	沿海城市 Coastal City	地区代码 Zip Code	沿海地带 Coastal County (District)	地区代码 Zip Code
		深圳 Shenzhen	440300	罗湖区Luohu Qu	440303
				福田区Futian Qu	440304
				南山区Nanshan Qu	440305
				宝安区Bao'an Qu	440306
				龙岗区Longgang Qu	440307
				盐田区Yantian Qu	440308
		珠海 Zhuhai	440400	香洲区Xiangzhou Qu	440402
				斗门区Doumen Qu	440403
				金湾区Jinwan Qu	440404
		汕头 Shantou	440500	龙湖区Longhu Qu	440507
				金平区Jinping Qu	440511
				濠江区Haojiang Qu	440512
				潮阳区Chaoyang Qu	440513
				潮南区Chaonan Qu	440514
				澄海区Chenghai Qu	440583
				南澳县Nan'ao Xian	440523
		江门 Jiangmen	440700	蓬江区Pengjiang Qu	440703
				江海区Jianghai Qu	440704
				新会区Xinhui Qu	440705
				台山市Taishan Shi	440781
				恩平市Enping Shi	440785
		湛江 Zhanjiang	440800	赤坎区Chikan Qu	440802
				霞山区Xiashan Qu	440803
				坡头区Potou Qu	440804
				麻章区Mazhang Qu	440811
				遂溪县Suixi Xian	440823
				徐闻县Xuwen Xian	440825
				廉江市Lianjiang Shi	440881
				雷州市Leizhou Shi	440082
				吴川市Wuchuan Shi	440083
		茂名 Maoming	440900	茂南区Maonan Qu	440902
				茂港区Maogang Qu	440903
				电白县Dianbai Xian	440923
		惠州 Huizhou	441300	惠城区Huicheng Qu	441302
				惠阳区Huiyang Qu	441303
				惠东县Huidong Xian	441323

1-2 续表6 continued

沿海地区 Coastal Region	地区代码 Zip Code	沿海城市 Coastal City	地区代码 Zip Code	沿海地带 Coastal County (District)	地区代码 Zip Code
		汕尾 Shanwei	441500	城　区Chengqu	441502
				海丰县Haifeng Xian	441521
				陆丰市Lufeng Shi	441581
		阳江 Yangjiang	441700	江城区Jiangcheng Qu	441702
				阳西县Yangxi Xian	441721
				阳东县Yangdong Xian	441723
		东莞 Dongguan	441900		
		中山 Zhongshan	442000		
		潮州 Chaozhou	445100	湘桥区Xiangqiao Qu	445102
				饶平县Raoping Xian	445122
		揭阳 Jieyang	445200	榕城区Rongcheng Qu	445202
				揭东县Jiedong Xian	445221
				惠来县Huilai Xian	445224
广西 Guangxi	450000	北海 Beihai	450500	海城区Haicheng Qu	450502
				银海区Yinhai Qu	450503
				铁山港区Tieshangang Qu	450512
				合浦县Hepu Xian	450521
		防城港 Fangchenggang	450600	港口区Gangkou Qu	450602
				防城区Fangcheng Qu	450603
				东兴市Dongxing Shi	450681
		钦州 Qinzhou	450700	钦南区Qinnan Qu	450702
海南 Hainan	460000	海口 Haikou	460100	秀英区Xiuying Qu	460105
				龙华区Longhua Qu	460106
				美兰区Meilan Qu	460108
		三亚 Sanya	460200		
				琼海市 Qionghai Shi	469002
				儋州市 Danzhou Shi	469003
				文昌市 Wenchang Shi	469005
				万宁市 Wanning Shi	469006
				东方市 Dongfang Shi	469007
				澄迈县Chengmai Xian	469023
				临高县Lingao Xian	469024
				昌江黎族自治县 Changjiang Lizu Zizhixian	469026
				乐东黎族自治县 Ledong Lizu Zizhixian	469027
				陵水黎族自治县 Lingshui Lizu Zizhixian	469028

1-3 海洋自然地理
Marine Physical Geography

指　　标		Item	指标值 Data
海洋平均深度	（米）	Average Depth of Sea　(m)	961
海洋最大深度	（米）	Maximum Depth of Sea　(m)	5 377
岸线总长度	（公里）	Length of Coastline　(km)	32 000
大陆岸线长度		Mainland Shore	18 000
岛屿岸线长度		Island Shore	14 000
岛屿个数	（个）	Number of Islands (unit)	5 400
岛屿面积	（万平方公里）	Area of Islands (10 000 km^2)	3.87

。

Note: The data come from the *China Statistical Yearbook 2010*（The same as in the Table 1-4）. Island area does not include that of Hong Kong Special Administrative Region, Macao Specical Administrative Region and Taiwan Province.

1-4 海区海域面积
Sea Area

自然海区名称 Natural Sea Area		海域总面积（千公顷） Sea Area (1 000 hm^2)	平均深度（米） Average Depth (m)	最大深度（米） Maximum Depth (m)
合　计	**Total**	**472 700**		
渤　海	Bohai Sea	7 700	18	70
黄　海	Huanghai Sea	38 000	44	140
东　海	Donghai Sea	77 000	370	2 719
南　海	Nanhai Sea	350 000	1 212	5 559

1-5 海洋自然资源
Ocean Natural Resources

指　　标	Item	指标值 Data
海洋能源理论蕴藏量　（亿千瓦）	Theoretical Sea-energy Reserves (100 million kW)	6.3
海水可养殖面积　（万公顷）	Cultivatable Area in Marine Areas (10 000 hm^2)	260.01
#已养殖面积	Cultivated Area	109.49
浅海滩涂可养殖面积　（万公顷）	Cultivable Area in Shallow Sea and Sea-beaches (10 000 hm^2)	242.00
#已养殖面积	Cultivated Area	89.37

注：数据来源于《2007中国统计年鉴》。

Note: The data come from the *China Statistical Yearbook 2007*.

1-6 海区海洋石油储量
Offshore Oil Reserves in the Sea Area

自然海区名称 Natural Sea Area	海洋石油（万吨） Offshore Oil (10 000 tons)	
	累计探明技术可采储量 Proven Technically Recoverable Reserves in the Aggregate	剩余技术可采储量 Surplus Technically Recoverable Reserves
合 计 Total	**74 563.4**	**40 164.5**
渤 海 Bohai Sea	42 289.5	29 299.5
黄 海 Huanghai Sea		
东 海 Donghai Sea	1 221.7	835.0
南 海 Nanhai Sea	31 052.2	10 030.0

注：数据来源于《2009年全国矿产资源储量通报》。

Note: The data come from the *Journal on the National Mineral Resources Reserves in 2009*.

1-7 沿海地区水资源基本情况
Water Resources by Coastal Regions

地 区 Region	水资源总量（亿立方米） Total Amount of Water Resources (100 million m^3)	地表水资源量 Surface Water Resources	地下水资源量 Groundwater Resources	地表水与地下水资源重复量 Duplicated Measurement Between Surface Water and Groundwater	人均水资源量（立方米/人） Per Capita Water Resources (m^3/person)
全国总计 National Total	**24 180.2**	**23 125.2**	**7 267.0**	**6 212.1**	**1 816.2**
天 津 Tianjin	15.2	10.6	5.6	0.9	126.8
河 北 Hebei	141.2	47.5	122.7	29.1	201.3
辽 宁 Liaoning	171.0	138.0	87.6	54.6	396.0
上 海 Shanghai	41.6	34.6	9.9	3.0	218.3
江 苏 Jiangsu	400.3	306.0	110.8	16.5	519.8
浙 江 Zhejiang	931.3	917.4	208.0	194.1	1 808.4
福 建 Fujian	800.8	799.6	244.7	243.4	2 214.9
山 东 Shandong	285.0	173.8	180.7	69.5	301.7
广 东 Guangdong	1 613.7	1 604.1	407.6	398.0	1 682.5
广 西 Guangxi	1 484.3	1 484.3	256.8	256.8	3 069.3
海 南 Hainan	480.7	474.6	106.3	100.3	5 596.2

注： 数据来源于《2010中国统计年鉴》。

Note: The data come from the *China Statistical Yearbook 2010*.

1-8 沿海地区湿地面积
Area of Wetlands by Coastal Regions

地 区 Region	湿地面积（千公顷）Area of Wetlands (1 000 hm^2)	天然湿地 Natural Wetlands	近岸及海岸 Coasts and Seashores	湿地面积占国土面积比重（%）Proportion of Wetlands in Total Area of Territory (%)
全国总计 National Total	**38 485.5**	**36 200.6**	**5 941.7**	**4.01**
天 津 Tianjin	171.8	133.7	58.1	14.95
河 北 Hebei	1 081.9	1 042.3	278.8	5.82
辽 宁 Liaoning	1 219.6	1 106.8	738.1	8.37
上 海 Shanghai	319.7	319.4	305.4	53.68
江 苏 Jiangsu	1 674.7	1 651.1	843.5	16.32
浙 江 Zhejiang	802.2	695.9	574.3	7.88
福 建 Fujian	443.0	421.2	370.6	3.65
山 东 Shandong	1 784.1	1 681.4	1 210.9	11.72
广 东 Guangdong	1 398.1	1 252.0	1 017.8	7.86
广 西 Guangxi	656.1	567.5	348.4	2.76
海 南 Hainan	311.5	256.6	190.0	9.13

注：数据来源于《2010中国统计年鉴》。本表为中国首次湿地调查(1995～2003)资料，不包括台湾省、香港和澳门特别行政区；湿地面积不包括水稻田湿地。

Note: The data come from the *China Statistical Yearbook 2010*. Data in the table are the figures of China First Wetlands Survey (1995-2003), excluding the wetlands of Taiwan province, Hong Kong SAR and Macao SAR. Area of wetlands excludes the wetland of paddyfield.

1-9 红树林各地类面积
Site Classification and Area of Sharpleaf Mangrove (Rhizophora Apiculata)

单位：公顷 (hm²)

地 区 Region	红树林各地类总面积 Total Site Area of Sharpleaf Mangrove	现有面积 Established	未成林面积 Unestablished	宜林地面积 Suitable for Planting
全国总计 National Total	**82 757.2**	**22 024.9**	**1 884.1**	**58 848.2**
浙 江 Zhejiang	5 452.3	20.6	236.1	5 195.6
福 建 Fujian	13 410.1	615.1	286.4	12 508.6
广 东 Guangdong	32 325.9	9 084.0	981.3	22 260.6
广 西 Guangxi	18 029.2	8 374.9	380.3	9 274.0
海 南 Hainan	13 539.7	3 930.3		9 609.4

注：数据来源于《2010中国统计年鉴》。本表数据为2002年全国红树林资源调查资料。

Note: The data come from the *China Statistical Yearbook 2010*. Data in the table are the figures of National Sharpleaf Mangrove Survey in 2002.

1-10 主要沿海城市气候基本情况
Climate of Major Coastal Cities

城 市 City	年平均气温 （摄氏度） Annual Average Temperature(℃)	年平均相对湿度 （%） Annual Average Relative Humidity (%)	全年降水量 （毫米） Annual Precipitation (millimeter)	全年日照时数 （小时） Annual Sunshine Hours (hour)
天 津 Tianjin	12.9	58	566.2	2 356.5
上 海 Shanghai	17.4	70	1 289.4	1 680.9
杭 州 Hangzhou	17.8	71	1 453.9	1 709.9
福 州 Fuzhou	20.7	70	1 374.7	1 605.6
广 州 Guangzhou	23.0	70	1 472.6	1 671.8
海 口 Haikou	24.3	81	2 628.2	1 861.1

注：数据来源于《2010中国统计年鉴》。

Note: The data come from the *China Statistical Yearbook 2010*.

主要统计指标解释

1. 沿海地区　即广义的沿海地区是指有海岸线（大陆岸线和岛屿岸线）的地区，按行政区划分为沿海省、自治区、直辖市。

2. 沿海城市　是指有海岸线的直辖市和地级市（包括其下属的全部区、县和县级市）。

3. 沿海地带　即狭义的沿海地区，是指有海岸线的县、县级市、区（包括直辖市和地级市的区）。

4. 海洋　是海和洋的统称。洋为地球表面上相连接的广大咸水水体的主体部分。海为地球表面相连接的广大咸水水体被陆地、岛礁、半岛包围或分隔的边缘部分。

5. 海水可养殖面积　指利用滩涂、浅海、港湾进行鱼、虾、蟹、贝、藻等海水经济动植物的人工养殖的水面面积。

6. 水资源总量　指评价区内降水形成的地表和地下产水总量，即地表产流量与降水入渗补给地下水量之和，不包括过境水量。

7. 地表水资源量　指评价区内河流、湖泊、冰川等地表水体中可以逐年更新的动态水量，即当地天然河川径流量。

8. 地下水资源量　指评价区内降水和地表水对饱水岩土层的补给量，包括降水入渗补给量和河道、湖库、渠系、渠灌田间等地表水体的入渗补给量。

9. 地表水与地下水资源重复量　指地表水和地下水相互转化的部分，即天然河川径流量中的地下水排泄量和地下水补给量中来源于地表水的入渗补给量。

10. 湿地　指天然或人工、长久或暂时性的沼泽地、泥炭地或水域地带，包括静止或流动、淡水、半咸水、咸水体，低潮时水深不超过6米的水域以及海岸地带地区的珊瑚滩和海草床、滩涂、红树林、河口、河流、淡水沼泽、沼泽森林、湖泊、盐沼及盐湖。

11. 红树林　指生长在热带、亚热带低能海岸潮间带上部，受周期性潮水浸淹，以红树植物为主体的常绿灌木或乔木组成的潮滩湿地木本生物群落。

12. 气温　指空气的温度，我国一般以摄氏度(℃)为单位表示。气象观测的温度表是放在离地面约1.5米处通风良好的百叶箱里测量的，因此，通常说的气温指的是离地面1.5米处百叶箱中的温度。其统计计算方法为：

月平均气温是将全月各日的平均气温相加，除以该月的天数而得。

年平均气温是将12个月的月平均气温累加后除以12而得。

13. 相对湿度　指空气中实际所含水蒸气密度和同温度下饱和水蒸气密度的百分比值。其统计方法与气温相同。

14. 降水量　指从天空降落到地面的液态或固态(经融化后)水，未经蒸发、渗透、流失而在地面上积聚的深度。其统计计算方法为：

月降水量是将全月各日的降水量累加而得。

年降水量是将12个月的月降水量累加而得。

15. 日照时数　指太阳实际照射地面的时间。其统计方法与降水量相同。

Explanatory Notes on Main Statistical Indicators

1. Coastal Region, i.e., the coastal region in a broad sense, refers to the regions with coastlines (continental and island coastlines), which are divided into the coastal provinces, autonomous regions and municipalities directly under the Central Government according to the administrative zoning.

2. Coastal City refers to the municipalities directly under the Central Government and the prefecture-level cities (including all the districts, counties and county-level cities under them).

3. Coastal Zone, i.e., the coastal region in a narrow sense, refers to the counties, county-level cities and districts with coastlines (including the districts under the municipalities directly under the Central Government and the prefecture-level districts).

4. Ocean is the general name for sea and ocean. Ocean refers to the main body of large salt water connected with the earth surface . Sea refers to the edge areas of the salt water on the earth surface that are compartmentalized or surrounded by land, island, reef or peninsula.

5. Marine Cultivatable Areas refer to water areas in beach, shallow sea and bays that are used to breed marine cash propagation, such as fish, shrimp, crab, shellfish, alga and so on.

6. Total Water Resources refers to total volume of water resources measured as run-off for surface water from rainfall and recharge for groundwater in a given area, excluding transit water.

7. Surface Water Resources refers to total renewable resources which exist in rivers, lakes, glaciers and other collectors from rainfall and are measured as run-off of rivers.

8. Groundwater Resources refers to replenishment of aquifers with rainfall and surface water.

9. Duplicated Measurement between Surface Water and Groundwater refers to the exchange between surface water and groundwater, i.e. run-off of rivers includes some depletion into groundwater while groundwater includes some replenishment from surface water.

10. Wetlands refer to marshland and peat bog, whether natural or man-made, permanent or temporary; water covered areas, whether stagnant or flowing, with fresh or brackish-fresh or salty water that is less than 6 meters deep at low tide; as well as coral beach, weed beach, mud beach, mangrove, river outlet, rivers, fresh-water marshland, marshland forests, lakes, salty bog and salt lakes along the coastal areas.

11. Mangrove refers to evergreen woody plants or plant communities in tropical or sub-tropical zones which live between the sea and the land in areas which are inundated by tides.

12. Temperature refers to the air temperature. China uses centigrade as the unit. The thermometry used for weather observation is put in a breezy shutter, which is 1.5 meters high from the ground. Therefore, the commonly used temperature refers to the temperature in the breezy shutter 1.5 meters away from the ground. The calculation method is as follows:

Monthly average temperature is the summation of average daily temperature of one month divided by the actual days of that particular month.

Annual average temperature is the summation of monthly averages of a year divided by 12 months.

13. Relative Humidity refers to the ratio of actual water vapour pressure to the saturated water vapour density under the current temperature. The calculation method is the same as that of temperature.

14. Volume of Precipitation refers to the deepness of liquid state or solid state (thawed) water falling from the sky to the ground that has not evaporated, infiltrated or run off. The calculation method is as follows:

Monthly precipitation is the summation of daily precipitation of a month.

Annual precipitation is the summation of 12 months precipitation of a year.

15. Sunshine Hours refer to the actual hours of sun irradiating the earth. The calculation method is the same as that of the precipitation.

2 海洋经济核算

Marine Economic Accounting

2-1 全国海洋生产总值
National Gross Ocean Product

年 份 Year	海洋生产总值（亿元） Gross Ocean Product (100 million yuan)	第一产业 Primary Industry	第二产业 Secondary Industry	第三产业 Tertiary Industry	海洋生产总值占国内生产总值比重（%） Proportion of the Gross Ocean Product in GDP (%)	海洋生产总值增长速度（%） Growth Rate of the Gross Ocean Product (%)
2001	9 518.4	646.3	4 152.1	4 720.1	8.68	
2002	11 270.5	730.0	4 866.2	5 674.3	9.37	19.8
2003	11 952.3	766.2	5 367.6	5 818.5	8.80	4.2
2004	14 662.0	851.0	6 662.8	7 148.2	9.17	16.9
2005	17 655.6	1 008.9	8 046.9	8 599.8	9.64	16.3
2006	21 260.4	1 238.6	9 693.1	10 328.7	10.03	16.8
2007	25 073.0	1 377.5	11 361.8	12 333.8	9.74	14.2
2008	29 718.0	1 694.3	13 735.3	14 288.4	9.46	9.8
2009	32 277.6	1 857.7	14 980.3	15 439.5	9.47	9.2

2-2 全国海洋生产总值构成
Composition of National Gross Ocean Product

年 份 Year	第一产业 Primary Industry （%）	第二产业 Secondary Industry （%）	第三产业 Tertiary Industry （%）
2001	6.8	43.6	49.6
2002	6.5	43.2	50.3
2003	6.4	44.9	48.7
2004	5.8	45.4	48.8
2005	5.7	45.6	48.7
2006	5.8	45.6	48.6
2007	5.5	45.3	49.2
2008	5.7	46.2	48.1
2009	5.8	46.4	47.8

2-3 海洋及相关产业增加值
Added Values of Marine and Related Industries

单位：亿元 (100 million yuan)

年份 Year	合计 Total	海洋产业 Marine Industry	主要海洋产业 Major Marine Industry	海洋科研教育管理服务业 Industries of Marine Scientific Research, Education, Management and Service	海洋相关产业 Ocean-related Industries
2001	9 518.4	5 733.6	3 856.6	1 877.0	3 784.8
2002	11 270.5	6 787.3	4 696.8	2 090.5	4 483.2
2003	11 952.3	7 137.7	4 754.4	2 383.3	4 814.6
2004	14 662.0	8 710.1	5 827.7	2 882.5	5 951.9
2005	17 655.6	10 539.0	7 188.0	3 350.9	7 116.6
2006	21 260.4	12 622.2	8 817.2	3 805.0	8 638.2
2007	25 073.0	14 902.0	10 465.1	4 436.9	10 171.0
2008	29 718.0	17 591.2	12 176.0	5 415.2	12 126.8
2009	32 277.6	18 822.0	12 843.6	5 978.4	13 455.6

2-4 海洋及相关产业增加值构成
Composition of the Added Value of Marine and Related Industries

单位：% (%)

年份 Year	合计 Total	海洋产业 Marine Industry	主要海洋产业 Major Marine Industry	海洋科研教育管理服务业 Industries of Marine Scientific Research, Education, Management and Service	海洋相关产业 Ocean-related Industries
2001	100.0	60.2	40.5	19.7	39.8
2002	100.0	60.2	41.7	18.5	39.8
2003	100.0	59.7	39.8	19.9	40.3
2004	100.0	59.4	39.7	19.7	40.6
2005	100.0	59.7	40.7	19.0	40.3
2006	100.0	59.4	41.5	17.9	40.6
2007	100.0	59.4	41.7	17.7	40.6
2008	100.0	59.2	41.0	18.2	40.8
2009	100.0	58.3	39.8	18.5	41.7

2-5 全国主要海洋产业增加值
Gross Output Value and Added Value of Major Marine Industries

主要海洋产业 Major Marine Industry	增加值 （亿元） Added Value (100 million yuan)	比上年增长（%） (按可比价计算) Percentage of Increase Over Last Year (%) (at comparable price)
合 计 **Total**	**12 843.6**	**9.5**
海洋渔业 Marine Fishery Industry	2 440.8	10.7
海洋油气业 Offshore Oil and Natural Gas Industry	614.1	-10.4
海洋矿业 Marine Mining Industry	41.6	33.7
海洋盐业 Sea Salt Industry	43.6	0.9
海洋船舶工业 Marine Shipbuilding Industry	986.5	32.6
海洋化工业 Marine Chemical Industry	465.3	24.8
海洋生物医药业 Marine Biomedicine Industry	52.1	-8.3
海洋工程建筑业 Marine Engineering Architecture	672.3	71.4
海洋电力业 Marine Electric Power Industry	20.8	79.8
海水利用业 Marine Seawater Utilization Industry	7.8	2.5
海洋交通运输业 Maritime Communications and Transportation Industry	3 146.6	-8.0
滨海旅游业 Coastal Tourism	4 352.3	16.4

2-6 海洋渔业增加值
Added Value of Marine Fishery Industry

单位：亿元 (100 million yuan)

年 份 Year	增加值 Added Value
2001	966.0
2002	1 091.2
2003	1 145.0
2004	1 271.2
2005	1 507.6
2006	1 708.1
2007	1 910.0
2008	2 228.6
2009	2 440.8

2-7 海洋油气业增加值
Added Value of Offshore Oil and Gas Industry

单位：亿元 (100 million yuan)

年 份 Year	增加值 Added Value
2001	176.8
2002	181.8
2003	257.0
2004	345.1
2005	528.2
2006	668.9
2007	691.6
2008	1 020.5
2009	614.1

2-8 海洋矿业增加值
Added Value of Marine Mining Industry

单位：亿元　(100 million yuan)

年 份 Year	增加值 Added Value
2001	1. 0
2002	1. 9
2003	3. 1
2004	7. 9
2005	8. 3
2006	6. 6
2007	7. 2
2008	35. 2
2009	41. 6

注:2008年-2009年数据中部分地区统计矿种增加。

Note: The data for 2008 and 2009 include added kinds of minerals in some regions.

2-9 海洋盐业增加值
Added Value of Marine Salt Industry

单位：亿元　(100 million yuan)

年 份 Year	增加值 Added Value
2001	32. 6
2002	34. 2
2003	28. 4
2004	39. 0
2005	39. 1
2006	40. 9
2007	47. 5
2008	43. 6
2009	43. 6

2-10 海洋船舶工业增加值
Added Value of Marine Shipbuilding Industry

单位：亿元 (100 million yuan)

年　份 Year	增加值 Added Value
2001	109.3
2002	117.4
2003	152.8
2004	204.1
2005	275.5
2006	380.6
2007	550.0
2008	742.6
2009	986.5

2-11 海洋化工业增加值
Added Value of Marine Chemical Industry

单位：亿元 (100 million yuan)

年　份 Year	增加值 Added Value
2001	64.7
2002	77.1
2003	96.3
2004	151.5
2005	153.3
2006	187.2
2007	234.7
2008	416.8
2009	465.3

注:2008年-2009年数据中部分地区统计产品品种增加。

Note: The data for 2008 and 2009 include added kinds of statistical products in some regions.

2-12 海洋生物医药业增加值
Added Value of Marine Biomedicine Industry

单位：亿元 (100 million yuan)

年 份 Year	增加值 Added Value
2001	5.7
2002	13.2
2003	16.5
2004	19.0
2005	28.6
2006	28.3
2007	43.7
2008	56.6
2009	52.1

2-13 海洋工程建筑业增加值
Added Value of Marine Engineering Architecture

单位：亿元 (100 million yuan)

年 份 Year	增加值 Added Value
2001	109.2
2002	145.4
2003	192.6
2004	231.8
2005	257.2
2006	327.2
2007	392.5
2008	347.8
2009	672.3

2-14 海洋电力业增加值
Added Value of Marine Electric Power Industry

单位：亿元 (100 million yuan)

年　份 Year	增加值 Added Value
2001	1.8
2002	2.2
2003	2.8
2004	3.1
2005	3.5
2006	4.4
2007	5.1
2008	11.3
2009	20.8

2-15 海水利用业增加值
Added Value of Seawater Utilization Industry

单位：亿元 (100 million yuan)

年　份 Year	增加值 Added Value
2001	1.1
2002	1.3
2003	1.7
2004	2.4
2005	3.0
2006	3.5
2007	4.2
2008	7.4
2009	7.8

2-16 海洋交通运输业增加值
Added Value of Marine Communications and Transportation Industry

单位：亿元 (100 million yuan)

年　份 Year	增加值 Added Value
2001	1 316.4
2002	1 507.4
2003	1 752.5
2004	2 030.7
2005	2 373.3
2006	2 842.1
2007	3 353.1
2008	3 499.3
2009	3 146.6

2-17 滨海旅游业增加值
Added Value of Coastal Tourism

单位：亿元 (100 million yuan)

年　份 Year	增加值 Added Value
2001	1 072.0
2002	1 523.7
2003	1 105.8
2004	1 522.0
2005	2 010.6
2006	2 619.6
2007	3 225.8
2008	3 766.4
2009	4 352.3

2-18 沿海地区海洋生产总值
Gross Ocean Product by Coastal Regions

地 区 Region	海洋生产总值（亿元） Gross Ocean Product (100 million yuan)	第一产业 Primary Industry	第二产业 Secondary Industry	第三产业 Tertiary Industry	海洋生产总值占沿海地区生产总值比重（%） Proportion of the Gross Ocean Product in the Gross Regional Product（%）
合 计 Total	**32 277.6**	**1 857.7**	**14 980.3**	**15 439.5**	**15.6**
天 津 Tianjin	2 158.1	5.1	1 329.3	823.6	28.7
河 北 Hebei	922.9	37.1	503.4	382.4	5.4
辽 宁 Liaoning	2 281.2	330.8	982.8	967.6	15.0
上 海 Shanghai	4 204.5	3.8	1 660.4	2 540.3	27.9
江 苏 Jiangsu	2 717.4	169.5	1 403.5	1 144.5	7.9
浙 江 Zhejiang	3 392.6	238.3	1 558.9	1 595.3	14.8
福 建 Fujian	3 202.9	272.1	1 408.9	1 521.9	26.2
山 东 Shandong	5 820.0	406.6	2 890.8	2 522.6	17.2
广 东 Guangdong	6 661.0	184.4	2 971.7	3 504.9	16.9
广 西 Guangxi	443.8	94.0	167.5	182.4	5.7
海 南 Hainan	473.3	116.1	103.1	254.1	28.6

2-19 沿海地区海洋生产总值构成
Composition of Gross Ocean Product by Coastal Regions

单位：%　　(%)

地 区 Region	海洋生产总值 Gross Ocean Product	第一产业 Primary Industry	第二产业 Secondary Industry	第三产业 Tertiary Industry
合 计 Total	**100.0**	**5.8**	**46.4**	**47.8**
天 津 Tianjin	100.0	0.2	61.6	38.2
河 北 Hebei	100.0	4.0	54.5	41.4
辽 宁 Liaoning	100.0	14.5	43.1	42.4
上 海 Shanghai	100.0	0.1	39.5	60.4
江 苏 Jiangsu	100.0	6.2	51.6	42.1
浙 江 Zhejiang	100.0	7.0	46.0	47.0
福 建 Fujian	100.0	8.5	44.0	47.5
山 东 Shandong	100.0	7.0	49.7	43.3
广 东 Guangdong	100.0	2.8	44.6	52.6
广 西 Guangxi	100.0	21.2	37.7	41.1
海 南 Hainan	100.0	24.5	21.8	53.7

2-20 沿海地区海洋及相关产业增加值
Added Values of Marine and Related Industries by Coastal Regions

单位：亿元 (100 million yuan)

地 区 Region	合 计 Total	海洋产业 Marine Industry	主要海洋产业 Major Marine Industry	海洋科研教育管理服务业 Industries of Marine Scientific Research, Education, Management and Service	海洋相关产业 Ocean-related Industries
合 计 Total	**32 277.6**	**18 822.0**	**12 843.6**	**5 978.4**	**13 455.6**
天 津 Tianjin	2 158.1	1 159.4	1 035.6	123.7	998.7
河 北 Hebei	922.9	492.5	430.5	61.9	430.4
辽 宁 Liaoning	2 281.2	1 439.6	1 129.2	310.4	841.7
上 海 Shanghai	4 204.5	2 495.4	1 540.4	954.9	1 709.1
江 苏 Jiangsu	2 717.4	1 533.5	1 130.4	403.0	1 184.0
浙 江 Zhejiang	3 392.6	1 935.5	1 335.3	600.2	1 457.1
福 建 Fujian	3 202.9	1 718.2	1 231.1	487.1	1 484.7
山 东 Shandong	5 820.0	3 201.1	2 238.6	962.4	2 619.0
广 东 Guangdong	6 661.0	4 226.2	2 304.3	1 921.9	2 434.7
广 西 Guangxi	443.8	284.3	228.4	55.9	159.5
海 南 Hainan	473.3	336.5	239.6	96.9	136.8

2-21 沿海地区海洋及相关产业增加值构成
Composition of the Added Value of Marine and Related Industries by Coastal Regions

单位：%　　(%)

地 区 Region	合 计 Total	海洋产业 Marine Industry	主要海洋产业 Major Marine Industry	海洋科研教育管理服务业 Industries of Marine Scientific Research, Education, Management and Service	海洋相关产业 Ocean-related Industries
合 计 Total	**100.0**	**58.3**	**39.8**	**18.5**	**41.7**
天 津 Tianjin	100.0	53.7	48.0	5.7	46.3
河 北 Hebei	100.0	53.4	46.7	6.7	46.6
辽 宁 Liaoning	100.0	63.1	49.5	13.6	36.9
上 海 Shanghai	100.0	59.4	36.6	22.7	40.6
江 苏 Jiangsu	100.0	56.4	41.6	14.8	43.6
浙 江 Zhejiang	100.0	57.1	39.4	17.7	42.9
福 建 Fujian	100.0	53.6	38.4	15.2	46.4
山 东 Shandong	100.0	55.0	38.5	16.5	45.0
广 东 Guangdong	100.0	63.4	34.6	28.9	36.6
广 西 Guangxi	100.0	64.1	51.5	12.6	35.9
海 南 Hainan	100.0	71.1	50.6	20.5	28.9

主要统计指标解释

1. 海洋经济 是开发、利用和保护海洋的各类产业活动以及与之相关联活动的总和。

2. 海洋生产总值 是海洋经济生产总值的简称，指按市场价格计算的沿海地区常住单位在一定时期内海洋经济活动的最终成果，是海洋产业和海洋相关产业增加值之和。

3. 海洋产业 是开发、利用和保护海洋所进行的生产和服务活动，包括海洋渔业、海洋油气业、海洋矿业、海洋盐业、海洋化工业、海洋生物医药业、海洋电力业、海水利用业、海洋船舶工业、海洋工程建筑业、海洋交通运输业、滨海旅游业等主要海洋产业以及海洋科研教育管理服务业。

4. 海洋科研教育管理服务业 是开发、利用和保护海洋过程中所进行的科研、教育、管理及服务等活动，包括海洋信息服务业、海洋环境监测预报服务、海洋保险与社会保障业、海洋科学研究、海洋技术服务业、海洋地质勘查业、海洋环境保护业、海洋教育、海洋管理、海洋社会团体与国际组织等。

5. 海洋相关产业 是指以各种投入产出为联系纽带，与主要海洋产业构成技术经济联系的上下游产业，涉及海洋农林业、海洋设备制造业、涉海产品及材料制造业、涉海建筑与安装业、海洋批发与零售业、涉海服务业等。

6. 海洋三次产业 我国的海洋三次产业划分如下：

海洋第一产业：是指海洋渔业中的海洋水产品、海洋渔业服务业，以及海洋相关产业中属于第一产业范畴的部门。

海洋第二产业：是指海洋渔业中海洋水产品加工、海洋油气业、海洋矿业、海洋盐业、海洋化工业、海洋生物医药业、海洋电力业、海水利用业、海洋船舶工业、海洋工程建筑业，以及海洋相关产业中属于第二产业范畴的部门。

海洋第三产业：是指除海洋第一、二产业以外的其他行业。第三产业包括：海洋交通运输业、滨海旅游业、海洋科研教育管理服务业，以及海洋相关产业中属于第三产业范畴的部门。

7. 海洋渔业 包括海水养殖、海洋捕捞、海洋渔业服务业和海洋水产品加工等活动。

8. 海洋油气业 是指在海洋中勘探、开采、输送、加工原油和天然气的生产活动。

9. 海洋矿业 包括海滨砂矿、海滨土砂石、海滨地热与煤矿及深海矿物等的采选活动。

10. 海洋盐业 是指利用海水生产以氯化钠为主要成分的盐产品的活动，包括采盐和盐加工。

11. 海洋船舶工业 是指以金属或非金属为主要材料，制造海洋船舶、海上固定及浮动装置的活动，以及对海洋船舶的修理及拆卸活动。

12. 海洋化工业 包括海盐化工、海水化工、海藻化工及海洋石油化工的化工产品生产活动。

13. 海洋生物医药业 是指以海洋生物为原料或提取有效成分，进行海洋药品与海洋保健品的生产加工及制造活动。

14. 海洋工程建筑业 是指在海上、海底和海岸所进行的用于海洋生产、交通、娱乐、防护等用途的建筑工程施工及其准备活动；包括海港建筑、滨海电站建筑、海岸堤坝建筑、海洋隧道桥梁建筑、海上油气田陆地终端及处理设施建造、海底线路管道和设备安装，不包括各部门、各地区的房屋建筑及房屋装修工程。

15. 海洋电力业 是指在沿海地区利用海洋能、海洋风能进行的电力生产活动。不包括沿海地区的火力发电和核力发电。

16. 海水利用业 是指对海水的直接利用和海水淡化活动，包括利用海水进行淡水生产和将海水应用于工业冷却用水和城市生活用水、消防用水等活动，不包括海水化学资源综合利用活动。

17. 海洋交通运输业 是指以船舶为主要工具从事海洋运输以及为海洋运输提供服务的活动，包括远洋旅客运输、沿海旅客运输、远洋货物运输、沿海货物运输、水上运输辅助活动、管道运输业、装卸搬运及其他运输服务活动。

18. 滨海旅游业 是指以海岸带、海岛及海洋各种自然景观、人文景观为依托的旅游经营、服务活动，主要包括：海洋观光游览、休闲娱乐、度假住宿、体育运动等活动。

Explanatory Notes on Main Statistical Indicators

1. Marine Economy is the summation of various types of industrial activities for developing, utilizing and protecting the ocean as well as the activities associated with there.

2. Gross Ocean Product is the short form of the gross output value of ocean economy, referring to the final result of marine economic activities of the permanent units in the coastal region within a given period calculated at the market price, and the sum total of the added values of the marine industries as the ocean-related industries.

3. Marine industry refers to the production as service activities for developing, utilizing and protecting the ocean, including major marine industries such as offshore oil and gas industry, marine mining industry, marine salt industry, marine chemical industry, marine biomedicine industry, marine electric power industry, seawater utilization industry, marine shipbuilding industry, marine engineering construction industry, marine communications and transportation industry, coastal tourism etc. as well as marine scientific research, education, management and service.

4. Marine Scientific Research, Education, Management and Service refer to the activities of scientific research, education, management and service carried out in the process of developing, utilizing and protecting the ocean, including marine information service industry, marine environment monitoring and forecasting service, marine insurance and social security industry, marine scientific research, marine technological service industry, ocean geological prospecting industry, marine environmental protection industry, marine education, marine management ,marine social organization and international organizations etc.

5. Ocean-Related Industry refers to the lower and upper reaches enterprises that form a technical and economic link with the major marine industries, with various inputs and outputs as ties, involving marine agriculture and forestry, marine equipment manufacturing, ocean-related building and installation industry, marine wholesale and retail industry, ocean-related service industry etc.

6. Marine Three Industries Chinese marine three industries are divided as follows:

Marine primary industry: refers to the marine aquatic products, marine fishery service industry in the marine fishery as well as the sectors belonging to the primary industry category in the ocean-related industries.

Marine secondary industry: refers to the marine aquatic products processing industry in the marine fishery, offshore oil as gas industry, marine mining industry, marine salt industry, marine chemical industry, marine biomedicine industry, marine electric power industry, seawater utilization industry, marine shipbuilding industry, marine engineering construction industry, as well as the sectors belonging to the category of secondary industry in the ocean-related industries.

Marine tertiary Industry: refers to the industries other than the marine primary and secondary industries, including marine communications and transportation industry, coastal tourism, marine scientific research, education, management and service industry as well as the sectors belonging to the category of tertiary industry in the ocean-related industries.

7. Marine Fishery includes mariculture, marine fishing, marine fishery service industry and marine aquatic products processing, etc.

8. Offshore Oil and Gas Industry refers to the production activities of exploring, exploiting, transporting and processing crude oil and natural gas in the ocean.

9. Ocean Mining Industry includes the activities of extracting and dressing beach placers, beach soil and sand, submarine geothermal energy, and coal mining and deep-sea mining, etc.

10. Marine Salt Industry refers to the activity of producing the salt products with the sodium chloride as the main component by utilizing seawater, including salt extracting and processing.

11. Shipbuilding Industry refers to the activity of building ocean vessels, offshore fixed and floating equipment with metals or non-metals as main materials as well as repairing and dismantling ocean vessels.

12. Marine Chemical Industry includes the production activities of chemical products of sea salt, seawater, sea algal and marine petroleum chemical industries.

13. Marine Biomedicine Industry refers to the production, processing and manufacturing activities of marine medicines and marine health care products by using marine organisms as raw materials or extracting useful components therefrom.

14. Marine Engineering Building Industry refers to the architectural projects construction and its preparations in the sea, at the sea bottom and seacoast for such uses as marine production, transportation, recreation, protection, etc., including constructions of seaports, coastal power stations, coastal dykes, marine tunnels and bridges, land terminals of offshore oil and gas fields as well as building of processing facilities, and installation of submarine pipelines and equipment, but not the projects of house building and renovation.

15. Marine Electric Power Industry refers to the activities of generating electric power in the coastal region by making use of ocean energies and ocean wind energy. It does not include the thermal and nuclear power generation in the coastal area.

16. Seawater Utilization Industry refers to the activities of the direct use of sea water and the seawater desalination, including those of carrying out the production of desalination and applying the seawater as water for industrial cooling, urban domestic water, water for fire fighting etc., but not the activity of the multipurpose use of seawater chemical resources.

17. Marine Communications and Transportation Industry refers to the activities of carrying out and serving the sea transportations with vessels as main vehicles, including ocean-going passagers transportation, coastal passagers transportation, ocean-going cargo transportation, coastal cargo transportation, auxiliary activities of water transportation, pipeline transportation, loading, unloading and transport as well as other transportation service activities.

18. Coastal Tourism refers to the tourist business and service activities with the backing of coastal zone, sea islands as well as a variety of natural and human landscapes of the ocean, mainly including marine sightseeing, living a life of leisure and recreation, going on vocation and getting accommodation, sports, etc.

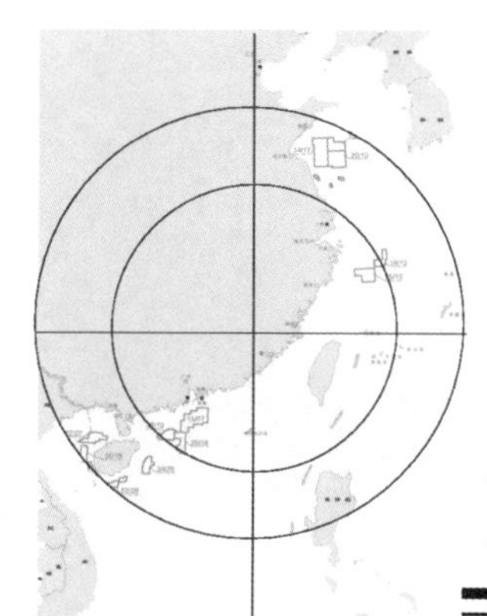

3

主要海洋产业活动

Major Marine Industrial Activities

3-1 全国海洋捕捞养殖产量
National Marine Catches and Mariculture Production

单位：吨 (ton)

项　目　Item	2007	2008	2009
海水水产品产量 Total Seawater Aquatic Products	**25 508 880**	**27 844 671**	**28 805 399**
海洋捕捞产量 Marine Catches	**12 435 480**	**13 408 617**	**13 440 772**
按品种分 By Species			
鱼类 Fish	8 224 231	9 086 773	9 148 657
甲壳类 Crustacea	2 070 402	2 175 080	2 177 651
贝类 Shellfish	743 617	660 647	681 832
藻类 Algae	32 847	36 643	28 542
头足类 Cephalopoda	1 047 713	942 980	820 410
其他 Others	316 670	376 154	438 366
海水养殖产量 Mariculture Production	**13 073 400**	**14 436 054**	**15 364 627**
按品种分 By Species			
鱼类 Fish	688 563	884 310	938 601
甲壳类 Crustacea	919 008	1 003 893	1 102 028
贝类 Shellfish	9 938 377	10 742 153	11 398 288
藻类 Algae	1 355 536	1 483 395	1 549 394
其他 Others	171 916	322 303	376 316

注：海洋捕捞产量分项中未包括辽宁远洋捕捞产量（2008，2009）。
Note: Data for Liaoning Deep-Sea fishing production is not included in the marine catches (2008, 2009).

3-2 沿海地区海洋捕捞养殖产量
Marine Catches and Mariculture Production by Coastal Regions

单位：吨 (ton)

地 区 Region	海洋捕捞产量 Marine Catches	# 远洋捕捞产量 Deep-Sea Fishing Production	海水养殖产量 Mariculture Production
合 计 **Total**	**13 440 772**	**818 569**	**15 364 627**
天 津 Tianjin	16 459	8 929	14 067
河 北 Hebei	253 317	1 197	300 567
辽 宁 Liaoning	1 483 097	138 079	2 896 175
上 海 Shanghai	168 500	146 657	0
江 苏 Jiangsu	570 008	7 345	734 960
浙 江 Zhejiang	3 152 295	160 473	857 893
福 建 Fujian	2 049 374	167 715	2 930 254
山 东 Shandong	2 370 891	78 700	3 814 304
广 东 Guangdong	1 525 341	109 474	2 346 157
广 西 Guangxi	801 088		1 271 630
海 南 Hainan	1 050 402		198 620

3-3 沿海地区海洋原油产量
Output of Offshore Crude Oil by Coastal Regions

单位：万吨 (10 000 tons)

地 区 Region	2007	2008	2009
合 计 Total	**3 178.37**	**3 421.13**	**3 698.19**
天 津 Tianjin	1 484.53	1 557.15	1 874.01
河 北 Hebei	164.32	189.62	200.30
辽 宁 Liaoning	23.17	22.80	15.00
上 海 Shanghai	21.56	15.34	9.68
山 东 Shandong	226.09	232.15	240.01
广 东 Guangdong	1 258.70	1 404.07	1 359.19

3-4 沿海地区海洋天然气产量
Output of Offshore Natural Gas by Coastal Regions

单位：万立方米 (10 000 m^3)

地 区 Region	2007	2008	2009
合 计 Total	**823 455**	**857 847**	**859 173**
天 津 Tianjin	160 686	140 111	143 002
河 北 Hebei	3 678	18 887	37 043
辽 宁 Liaoning	8 231	5 937	4 575
上 海 Shanghai	75 395	63 716	58 767
山 东 Shandong	15 563	16 798	16 157
广 东 Guangdong	559 902	612 398	599 629

3-5 海洋原油出口量及创汇额
Export Volume and Foreign-Exchange Earnings of Offshore Crude Oil by Coastal Regions

单位：万吨，万美元　　　　(10 000 tons, 10 000 US$)

地　区 Region	2007		2008		2009	
	出口量 Export Volume	创汇额 Foreign-Exchange Earnings	出口量 Export Volume	创汇额 Foreign-Exchange Earnings	出口量 Export Volume	创汇额 Foreign-Exchange Earnings
合　计 Total	**278.26**	**120 220**	**318.09**	**209 533**	**134.90**	**51 901**
天　津 Tianjin	187.90	77 049	186.02	113 629	66.15	24 679
广　东 Guangdong	90.36	43 171	132.07	95 904	68.75	27 222

3-6 海洋原油产量、出口量占全国原油产量、出口量比重
Proportion of Offshore Crude Oil Production and Export Volume in the National Total

年　份 Year	海洋原油产量占全国原油产量比重（%） Proportion of Offshore Crude Oil Production in the National Total (%)	海洋原油出口量占全国原油出口量比重（%） Proportion of Offshore Crude Oil Export Volume in the National Total (%)
2001	13.07	46.26
2002	14.40	54.95
2003	15.01	60.77
2004	16.16	83.37
2005	17.51	84.18
2006	17.54	89.53
2007	17.06	71.53
2008	18.00	76.46
2009	19.52	26.61

3-7 沿海地区海洋矿业产量
Output of Marine Mining Industry by Coastal Regions

单位：吨 (ton)

地 区 Region	产 量 Output		
	2007	2008	2009
合 计 Total	**29 581 575**	**48 085 689**	**55 906 609**
浙 江 Zhejiang	24 137 754	40 015 300	47 554 400
福 建 Fujian	2 030 700	2 063 800	2 082 500
山 东 Shandong	385 665	3 308 591	3 423 889
广 东 Guangdong	153 000		
广 西 Guangxi	1 266 456	887 998	564 820
海 南 Hainan	1 608 000	1 810 000	2 281 000

3-8 沿海地区海盐产量
Productive Activities of Salt-Making Industry by Coastal Regions

单位：万吨 (10 000 tons)

地区 Region	海盐产量 Output of Sea Salt		
	2007	2008	2009
合计 Total	**3 191.70**	**3 127.15**	**3 500.45**
天津 Tianjin	237.03	235.96	227.56
河北 Hebei	411.40	385.07	421.65
辽宁 Liaoning	220.26	182.08	222.33
江苏 Jiangsu	129.38	102.43	106.61
浙江 Zhejiang	29.05	19.28	17.05
福建 Fujian	42.34	40.13	53.14
山东 Shandong	2 070.91	2 122.71	2 412.72
广东 Guangdong	20.02	14.54	11.57
广西 Guangxi	13.64	13.18	15.34
海南 Hainan	17.67	11.77	12.48

3-9 沿海地区海洋化工产品产量
Output of Marine Chemical Products by Coastal Regions

单位：吨 (ton)

地 区 Region	产品产量 Output		
	2007	2008	2009
合 计 Total	**20 416 992**	**14 558 214**	**11 993 540**
天 津 Tianjin	2 245 533	1 986 324	1 640 000
河 北 Hebei	1 869 238	771 750	820 340
辽 宁 Liaoning	327 765	771 622	446 095
江 苏 Jiangsu	1 371 416	310 549	362 463
浙 江 Zhejiang	1 898 020	444 866	373 133
福 建 Fujian	3 721 244	4 354 899	269 840
山 东 Shandong	8 351 696	5 197 364	7 240 769
广 东 Guangdong	632 080	720 840	840 900

3-10 沿海地区海洋生物医药产品产量*
Production of Marine Biomedicine Industry by Coastal Regions

产品名称 Name	计量单位 Unit	产品产量 Output
海参肽营养素胶囊 Sea cucumber peptide nutrient capsule	箱　box	538 800.00
藻日康、巨藻盖 Zaorikang,Ju Zaogai	盒　case	233 500.00
硫酸软骨素 Sulphate Chondroitin	公斤　kg	318 652.00
螺旋藻片 Spirulina Pill	万粒　10 000 pellets	150 000.00
螺旋藻粉 Spirulina Powder	公斤　kg	100.00
螺旋藻 Spirulina	万瓶　10 000 bottles	113.00
螺旋藻胶囊 Spirulina Capsule	瓶　bottle	2 000.00
藻酸双脂钠 Alginic Acid Diadipose Sodium	万瓶　10 000 bottles	91.09
藻酸双酯钠片 Alginic Acid Diadipose Sodium Pill	万片　10 000 pills	395.00
藻酸双酯钠注射液 Alginic Acid Diadipose Sodium Injection	万盒　10 000 cases	9.00
鲨鱼肝油胶丸 Shark Liver Oil Pill	万粒　10 000 pellets	3 945.00
金枪鱼油胶丸 Tuna Oil Pill	万粒　10 000 pellets	1 722.00
卵磷脂胶丸 Lecithin Pill	万粒　10 000 pellets	812.00
鳕鱼肝油胶丸 Ling Liver Oil Pill	万粒　10 000 pellets	176.00
海藻植物胶囊 Algae Plant Capsule	万粒　10 000 pellets	8.50
蚝贝钙片 Oyster Shell Calcium Tablets	万片　10 000 pills	2 651.35
三维鱼肝油乳 Sanwei Liver Oil Milk	万瓶　10 000 bottles	166.56
银耳鱼肝油乳 Tremella Liver Oil Milk	万瓶　10 000 bottles	32.00
鱼肝油 Cod-liver Oil	万瓶　10 000 bottles	227.00
鱼肝油乳 Cod-Liver Oil Milk	吨　ton	15.79
鱼油软胶囊 Fish Oil Soft Capsule	万粒　10 000 pellets	375.00
鱼油胶丸 Fish Oil Capsule	吨　ton	40.00
精炼鱼油 Refineed fish Oil	吨　ton	1 530.00
多烯酸乙酯 Ethyl Polyenoic Acid	万粒　10 000 pellets	108.00
鲨鱼软骨胶囊 Shark Cartilage Capsule	万颗　10 000 pellets	44.00
鲨鱼硫酸骨素 Shark Sulphate Chondroitin	吨　ton	100.00
海蛇痹宁胶囊 Sea Snake Anti-rheumatic Capsule	万粒　10 000 pellets	267.75
维生素AD胶囊 Vitamin AD Capsule	万粒　10 000 pellets	31 724.19
维生素AD滴剂（贝特令） Vitamin AD drops	万粒　10 000 pellets	9 160.17
维生素VE胶囊 Vitamin VE Capsule	万粒　10 000 pellets	40 419.19
鲎试剂 King Crab Tablet	万支　10 000 bottles	346.00
海珠喘息定片 Haizhu Methoxyphenamine Pill	万片　10 000 pills	17 074.00
珍珠粉末 Pearl Powder	万瓶　10 000 bottles	620.00

3-10 续表 continued

产品名称 Name	计量单位 Unit	产品产量 Output
金牡感冒片 Jinmu Cold Cure Pill	万片 10 000 pills	1 029.32
片仔癀珍珠膏 Pianzaihuang pearl Ointment	万瓶 10 000 bottles	200.05
琼脂 Agar	吨 ton	258.00
碘[I-125]密封籽源 Iodine [I-125] sealed	粒 pellet	81 176.59
鳃腺炎减毒活疫苗 Parotitis toxicity reducing living vaccine	人/份 Person/portion	11 046 268.23
甲壳素 Crustaceoxin	吨 ton	2 758.00
海麒舒肝胶囊 Haiqi Liver-Soothing Capsule	万盒 10 000 cases	90.00
海墨止血片 Haimo Haemostatic Pill	万盒 10 000 cases	10.62
多烯酸乙酯胶囊 Ethyl Polyenoic Acid Capsule	万粒 10 000 pellets	1 071.83
多烯鱼油胶囊 Polyene Fish-oil capsule	万瓶 10 000 bottles	1.00
角鲨烯胶丸 Houndfish Alkene Capsule	万粒 10 000 pellets	941.37
鱼胶原蛋白肽 Fish Collagen Peptide	万包 10 000 packages	85.47
鱼蛋白粉 Fish Protein Powder	吨 ton	1.20
角鲨烯原料、多烯酸乙酯、保健品鱼油 Squalene raw material, polyene acid ethyl ester, health care fish-oil	吨 ton	1 591.60
胶原蛋白 Collagen	吨 ton	110.00
七叶皂甘钠 Qi Ye Zao Gan Na	万支 10 000 bottles	814.00
佰海先牌康体通胶囊 Baihaixian brand Kangtitong capsale	万粒 10 000 pellets	8 000.00
海参保健系列产品 Sea cucumber health care series products	盒 case	5 000.00
海威口服液 Haiwei oral liquid	吨 ton	430.00
脑元神软胶囊 Naoyuanshen Soft Capsule	万粒 10 000 pellets	31.35
依可新 Vitamin A and D Drops	吨 ton	198.00
乙酰胺基噻二唑 Acetazolamide	吨 ton	13.00
苯海因 Benzene Glycolylurea	吨 ton	1 360.00
甘糖酯片 Mannose Ester Tablets	万片 10 000 pills	36.00
氨曲南单环母核 An Qu Nan Dan Huan Mu He	吨 ton	11.00
谷胱苷肽 Glutathione	万支 10 000 bottles	1 016.00
微生物制剂 Microbological preparation	万箱 10 000 boxes	520.00
甲氧苄啶 Trimethoprim	万箱 10 000 boxes	300.00
羟混苯 Hydroxy Mixed Benzene	吨 ton	3 100.00
D酯 D ester	吨 ton	190.00
氨糖美辛肠溶片 An Tang Mei Xin Chang Rong Pian	万片 10 000 pills	1 099.22

注：数据为沿海地区部分海洋生物医药产品汇总数据。

Note: The data collected from the products of part of the marine biomedicine enterprises in the coastal region.

3-11 沿海地区海洋修造船完工量
Production of the Marine Shipbuilding Industry by Coastal Regions

部门和地区 Sector and Region	修船完工量（艘） Ships Repaired (number)	造船完工量 Ships Built	
		艘 number	万综合吨 Comprehensive Tonnages (10 000 tons)
总　计　Total	**8 982**	**1 979**	**4 439.31**
其　中: Including:			
中船工业集团公司 CSSC	421	124	1 013.60
中船重工集团公司 CSIC	623	70	557.05
按地区分: By Regions:			
天　津 Tianjin	220	24	29.00
河　北 Hebei	195	3	9.00
辽　宁 Liaoning	360	48	504.00
上　海 Shanghai	978	97	932.08
江　苏 Jiangsu	438	551	1 546.00
浙　江 Zhejiang	3 790	679	422.33
福　建 Fujian	1 761	325	93.92
山　东 Shandong	1 108	136	901.42
广　东 Guangdong			
广　西 Guangxi	16		0.63
海　南 Hainan	116	116	0.93

3-12 沿海地区海洋货物运输量和周转量
Maritime Volume of Goods Transported and Turnover by Coastal Regions

单位：万吨，亿吨公里 (10 000 tons, 100 million ton-km)

地区 Region	货运量 Volume of Goods Transported	沿海 Coastal	远洋 Oceangoing	货物周转量 Volume of Goods Turnover	沿海 Coastal	远洋 Oceangoing
全国总计 National Total	**162 164**	**110 431**	**51 733**	**52 923.93**	**13 399.81**	**39 524.12**
天津 Tianjin	11 256	2 045	9 211	8 942.54	247.27	8 695.27
河北 Hebei	1 052	1 052		223.82	223.82	
辽宁 Liaoning	9 636	4 661	4 975	4 896.82	595.56	4 301.26
上海 Shanghai	36 057	24 127	11 930	14 079.54	3 558.72	10 520.82
江苏 Jiangsu	11 795	7 921	3 874	2 733.61	791.51	1 942.10
浙江 Zhejiang	30 089	29 307	782	3 844.73	3 218.08	626.65
福建 Fujian	11 970	11 158	812	1 775.05	1 471.49	303.56
山东 Shandong	9 780	4 792	4 988	3 446.81	375.71	3 071.10
广东 Guangdong	16 686	9 778	6 908	2 701.45	1 382.27	1 319.18
广西 Guangxi	2 368	2 110	258	323.97	314.16	9.81
海南 Hainan	6 663	6 489	174	661.41	615.90	45.51
其他 Others	14 812	6 991	7 821	9 294.18	605.32	8 688.86

3-13 沿海地区海洋旅客运输量和周转量
Maritime Volume of Passenger Traffic and Turnover by Coastal Regions

单位：万人，亿人公里 (10 000 persons, 100 million person-km)

地 区 Region	客运量 Passenger Traffic	沿 海 Coastal	远 洋 Oceangoing	旅客周转量 Passenger Turnover Volume	沿 海 Coastal	远 洋 Oceangoing
全国总计 National Total	**10 918**	**10 121**	**797**	**41.80**	**33.57**	**8.23**
天 津 Tianjin	1		1	0.12		0.12
辽 宁 Liaoning	543	532	11	7.03	6.59	0.44
上 海 Shanghai	1 415	1 415		6.29	6.29	
江 苏 Jiangsu	18	18		0.78	0.78	
浙 江 Zhejiang	3 125	3 125		6.78	6.78	
福 建 Fujian	1 159	1 097	62	1.52	1.30	0.22
山 东 Shandong	2 043	1 977	66	9.81	7.03	2.78
广 东 Guangdong	1 612	956	656	6.54	1.89	4.65
广 西 Guangxi	94	93	1	0.62	0.60	0.02
海 南 Hainan	908	908		2.30	2.30	

3-14 沿海港口客货吞吐量
Passengers Leaving and Arriving and Cargo Handled at Coastal Seaports

单位：万吨，万人次 (10 000 tons, 10 000 person-times)

地 区 Region	货物吞吐量 Cargo Handled	# 外 贸 Foreign Trade	旅客吞吐量 Passenger Leaving & Arriving	# 离 港 Leaving
合 计 Total	**487 371**	**199 398**	**8 303**	**4 171**
天 津 Tianjin	38 111	19 633	16	8
河 北 Hebei	50 874	12 914	6	3
辽 宁 Liaoning	55 259	14 197	670	331
上 海 Shanghai	49 467	25 814	175	90
江 苏 Jiangsu	11 708	6 618	12	6
浙 江 Zhejiang	71 462	25 792	2 831	1 410
福 建 Fujian	30 542	10 519	764	381
山 东 Shandong	73 072	41 527	1 115	566
广 东 Guangdong	89 123	34 824	1 776	887
广 西 Guangxi	9 408	5 870	27	13
海 南 Hainan	8 345	1 690	911	478

3-15 沿海国际标准集装箱运量
Volume of International Standardized Containers Traffic at Coastal Seaports

单位：万标准箱，万吨 (10 000 TEU, 10 000 tons)

地 区 Region	2007		2008		2009	
	箱 数 Containers	重 量 Weight	箱 数 Containers	重 量 Weight	箱 数 Containers	重 量 Weight
合 计 Total	**2 953**	**33 356**	**3 219**	**34 975**	**3 011**	**34 578**
天 津 Tianjin	13	178	12	112	12	68
河 北 Hebei			2	15	1	20
辽 宁 Liaoning	35	334	16	146	39	393
上 海 Shanghai	1 593	19 420	1 511	18 418	1 332	17 710
江 苏 Jiangsu	211	1 862	190	1 670	331	2 702
浙 江 Zhejiang	48	862	53	669	80	1 529
福 建 Fujian	117	1 407	145	1 745	205	3 272
山 东 Shandong	131	1 664	170	1 746	132	1 416
广 东 Guangdong	647	5 666	795	6 772	704	5 063
广 西 Guangxi	17	165	145	1 209	49	675
海 南 Hainan	34	417	20	285	22	295
其 他 Others	107	1 381	160	2 188	106	1 435

注：国际标准集装箱运量数据包含内河集装箱运量。

Note: The data in this table include the volume of containers transported in the inland rivers.

3-16 沿海港口国际标准集装箱吞吐量
International Standardized Containers Handled at Coastal Seaports

单位：万标准箱，万吨　　(10 000TEU, 10 000 tons)

地　区 Region	2007		2008		2009	
	箱　数 Containers	重　量 Weight	箱　数 Containers	重　量 Weight	箱　数 Containers	重　量 Weight
合　计 Total	**10 470**	**99 528**	**11 673**	**114 692**	**11 020**	**114 415**
天　津 Tianjin	710	7 680	850	8 591	870	8 871
河　北 Hebei	49	683	65	888	57	869
辽　宁 Liaoning	582	7 652	744	10 372	812	11 718
上　海 Shanghai	2 615	23 850	2 801	25 992	2 500	24 619
江　苏 Jiangsu	201	2 004	301	2 816	305	2 872
浙　江 Zhejiang	987	7 855	1 148	9 357	1 118	9 839
福　建 Fujian	686	7 368	743	8 709	716	9 001
山　东 Shandong	1 165	11 586	1 321	14 052	1 312	13 817
广　东 Guangdong	3 407	29 851	3 620	32 779	3 236	31 432
广　西 Guangxi	27	431	34	523	35	555
海　南 Hainan	41	568	46	613	59	822

3-17 沿海城市国内旅游人数
Domestic Tourists by Coastal Cities

单位：万人次　　　　(10 000 person-times)

城　市	City	2006	2007	2008
合　计	**Total**	**51 423**	**65 875**	**74 700**
天　津	**Tianjin**		6 018	7 004
河　北	**Hebei**	2 377	2 613	2 553
唐　山	Tangshan	685	762	957
秦皇岛	Qinhuangdao	1 409	1 510	1 227
沧　州	Cangzhou	283	341	369
辽　宁	**Liaoning**	4 901	6 150	7 735
大　连	Dalian	2 150	2 480	3 000
丹　东	Dandong	950	1 100	1 420
锦　州	Jinzhou	532	680	870
营　口	Yingkou	285	450	585
盘　锦	Panjin	454	760	980
葫芦岛	Huludao	530	680	880
上　海	**Shanghai**	9 684	10 210	11 006
江　苏	**Jiangsu**	2 297	2 686	3 146
南　通	Nantong	883	1 072	1 275
连云港	Lianyungang	803	911	1 065
盐　城	Yancheng	611	703	806
浙　江	**Zhejiang**	14 565	16 879	19 229
杭　州	Hangzhou	3 694	4 112	4 552
宁　波	Ningbo	2 685	3 074	3 465
温　州	Wenzhou	1 843	2 192	2 547
嘉　兴	Jiaxing	1 595	1 850	2 140
绍　兴	Shaoxing	1 808	2 192	2 435
舟　山	Zhoushan	1 136	1 285	1 495
台　州	Taizhou	1 804	2 174	2 595

3-17 续表 continued

城　市 City		2006	2007	2008
山　东	**Shandong**	8 195	9 927	11 491
青　岛	Qingdao	2 801	3 259	3 390
东　营	Dongying	242	329	428
烟　台	Yantai	1 694	1 999	2 346
潍　坊	Weifang	1 038	1 414	1 869
威　海	Weihai	1 130	1 358	1 586
日　照	Rizhao	1 001	1 226	1 451
滨　州	Binzhou	289	342	421
广　东	**Guangdong**	8 253	9 238	10 170
广　州	Guangzhou	2 396	2 727	2 916
深　圳	Shenzhen	1 605	1 729	1 790
珠　海	Zhuhai	527	553	829
汕　头	Shantou	485	542	604
江　门	Jiangmen	572	619	728
湛　江	Zhanjiang	148	142	165
茂　名	Maoming	161	179	171
惠　州	Huizhou	461	586	675
汕　尾	Shanwei	162	189	219
阳　江	Yangjiang	237	238	240
东　莞	Dongguan	784	960	994
中　山	Zhongshan	421	442	463
潮　州	Chaozhou	169	193	219
揭　阳	Jieyang	125	139	157
广　西	**Guangxi**	373	1 097	1 191
北　海	Beihai		601	695
防城港	Fangchenggang	179	194	150
钦　州	Qinzhou	194	302	346
海　南	**Hainan**	778	1 057	1 175
海　口	Haikou	362	571	622
三　亚	Sanya	416	486	553

注：数据来源于《中国区域经济统计年鉴》（2009）。

Note：The data come from *China Statistical Yearbook For Regional Economy (2009)*.

3-18 沿海城市国际旅游（外汇）收入
Foreign Exchange Earnings from International Tourism by Coastal Cities

单位：万美元 (10 000 US$)

城 市	City	2007	2008	2009
合 计	**Total**	**2 037 088**	**2 189 709**	**2 514 037**
天 津	**Tianjin**	77 871	100 139	118 264
河 北	**Hebei**	14 962	12 035	14 414
唐 山	Tangshan	1 664	1 655	2 126
秦皇岛	Qinhuangdao	13 073	10 169	11 909
沧 州	Cangzhou	225	211	379
辽 宁	**Liaoning**	73 773	89 399	104 347
大 连	Dalian	58 125	65 835	72 748
丹 东	Dandong	5 927	9 040	12 535
锦 州	Jinzhou	4 530	6 512	8 317
营 口	Yingkou	1 178	2 159	2 640
盘 锦	Panjin	2 324	3 828	5 486
葫芦岛	Huludao	1 689	2 024	2 621
上 海	**Shanghai**	467 297	497 172	474 402
江 苏	**Jiangsu**	35 487	39 816	44 009
南 通	Nantong	24 728	28 368	30 933
连云港	Lianyungang	7 789	7 959	9 173
盐 城	Yancheng	2 970	3 490	3 903
浙 江	**Zhejiang**	217 604	243 411	254 784
杭 州	Hangzhou	112 665	129 610	137 995
宁 波	Ningbo	43 070	46 874	48 650
温 州	Wenzhou	14 083	16 109	17 797
嘉 兴	Jiaxing	19 080	18 917	19 182
绍 兴	Shaoxing	12 063	13 644	14 818
舟 山	Zhoushan	10 443	11 212	11 378
台 州	Taizhou	6 200	7 046	4 964

3-18 续表 continued

城市 City		2007	2008	2009
福建	**Fujian**	207 863	230 529	434 969
福州	Fuzhou	59 868	65 750	259 923
厦门	Xiamen	71 880	81 865	90 194
莆田	Putian	4 138	5 153	7 290
泉州	Quanzhou	49 398	65 980	64 771
漳州	Zhangzhou	22 469	11 653	12 685
宁德	Ningde	110	128	106
山东	**Shandong**	110 479	104 020	125 741
青岛	Qingdao	67 507	50 030	55 178
东营	Dongying	668	1 508	2 214
烟台	Yantai	22 950	26 708	31 081
潍坊	Weifang	3 607	7 081	12 254
威海	Weihai	12 447	13 734	16 083
日照	Rizhao	3 042	4 432	8 168
滨州	Binzhou	258	527	763
广东	**Guangdong**	800 274	840 098	916 884
广州	Guangzhou	319 147	313 035	362 396
深圳	Shenzhen	262 328	270 399	276 026
珠海	Zhuhai	90 240	94 823	102 670
汕头	Shantou	5 936	6 491	4 910
江门	Jiangmen	16 485	39 129	40 891
湛江	Zhanjiang	1 785	1 891	2 237
茂名	Maoming	986	1 099	981
惠州	Huizhou	28 669	33 091	40 107
汕尾	Shanwei	525	650	511
阳江	Yangjiang	2 126	1 537	1 412
东莞	Dongguan	42 702	45 614	51 756
中山	Zhongshan	21 678	22 702	20 434
潮州	Chaozhou	6 923	8 774	11 072
揭阳	Jieyang	744	862	1 481
广西	**Guangxi**	2 944	3 132	3 637
北海	Beihai	1 401	1 559	1 721
防城港	Fangchenggang	916	928	1 268
钦州	Qinzhou	627	645	648
海南	**Hainan**	28 534	29 957	22 586
海口	Haikou	4 126	3 702	3 089
三亚	Sanya	24 408	26 255	19 497

3-19 沿海城市接待入境旅游者人数

Number of Oversea Visitor Arrivals by Coastal Cities

单位：人次 (person-time)

城 市	City	2007	2008	2009
合 计	**Total**	**39 838 769**	**41 785 669**	**45 585 443**
天 津	**Tianjin**	1 032 268	1 220 392	1 410 244
河 北	**Hebei**	269 041	237 301	287 568
唐 山	Tangshan	40 631	42 934	50 414
秦皇岛	Qinhuangdao	221 828	187 267	224 206
沧 州	Cangzhou	6 582	7 100	12 948
辽 宁	**Liaoning**	1 190 030	1 422 628	1 688 798
大 连	Dalian	840 032	950 045	1 050 043
丹 东	Dandong	148 624	190 762	267 860
锦 州	Jinzhou	87 342	114 820	150 381
营 口	Yingkou	24 527	49 670	59 787
盘 锦	Panjin	53 365	82 110	115 000
葫芦岛	Huludao	36 140	35 221	45 727
上 海	**Shanghai**	5 200 981	5 264 727	5 333 935
江 苏	**Jiangsu**	354 594	421 663	454 880
南 通	Nantong	223 907	280 037	299 866
连云港	Lianyungang	82 204	90 922	100 076
盐 城	Yancheng	48 483	50 704	54 938
浙 江	**Zhejiang**	4 330 391	4 532 494	4 732 744
杭 州	Hangzhou	2 085 997	2 213 319	2 304 045
宁 波	Ningbo	689 231	756 776	800 548
温 州	Wenzhou	289 038	318 230	329 734
嘉 兴	Jiaxing	612 532	529 574	556 708
绍 兴	Shaoxing	360 915	398 811	431 713
舟 山	Zhoushan	199 565	211 964	223 482
台 州	Taizhou	93 113	103 820	86 514
福 建	**Fujian**	2 441 031	2 654 976	2 915 820
福 州	Fuzhou	589 373	639 375	629 314
厦 门	Xiamen	932 158	1 045 309	1 281 907

3-19 续表 continued

城 市 City		2007	2008	2009
莆 田	Putian	127 449	142 555	155 384
泉 州	Quanzhou	601 586	630 776	626 721
漳 州	Zhangzhou	186 738	192 709	219 382
宁 德	Ningde	3 727	4 252	3 112
山 东	**Shandong**	1 861 344	1 760 662	2 126 781
青 岛	Qingdao	1 080 341	800 836	1 000 670
东 营	Dongying	11 396	18 760	25 403
烟 台	Yantai	307 373	352 090	400 901
潍 坊	Weifang	74 159	131 771	172 530
威 海	Weihai	267 317	288 277	322 676
日 照	Rizhao	112 192	151 618	180 145
滨 州	Binzhou	8 566	17 310	24 456
广 东	**Guangdong**	22 360 364	23 506 096	26 075 166
广 州	Guangzhou	6 113 369	6 124 801	6 894 044
深 圳	Shenzhen	8 313 042	8 695 727	8 963 697
珠 海	Zhuhai	2 820 430	2 862 150	2 978 429
汕 头	Shantou	150 182	139 313	123 161
江 门	Jiangmen	575 890	1 113 981	2 340 828
湛 江	Zhanjiang	44 279	32 994	50 665
茂 名	Maoming	34 362	7 902	27 633
惠 州	Huizhou	1 212 749	1 327 919	1 442 523
汕 尾	Shanwei	25 725	24 930	36 640
阳 江	Yangjiang	66 293	54 717	43 668
东 莞	Dongguan	2 013 094	2 166 205	2 259 625
中 山	Zhongshan	682 562	651 753	475 769
潮 州	Chaozhou	271 066	262 900	389 664
揭 阳	Jieyang	37 321	40 804	48 820
广 西	**Guangxi**	127 253	117 126	138 468
北 海	Beihai	55 011	55 702	61 437
防城港	Fangchenggang	55 156	43 421	57 311
钦 州	Qinzhou	17 086	18 003	19 720
海 南	**Hainan**	671 472	647 604	421 039
海 口	Haikou	149 475	136 128	103 206
三 亚	Sanya	521 997	511 476	317 833

3-20 沿海城市接待入境旅游者情况
Breakdown of Inbound Tourists by Coastal Cities

单位：人次，人天 (person-time, night)

城市 City	外国人 Foreigners		香港同胞 Hong Kong	
	人次数 Arrivals	人天数 Nights	人次数 Arrivals	人天数 Nights
合　计 Total	**19 899 904**	**61 709 189**	**16 959 026**	**37 473 676**
天　津 **Tianjin**	1 305 772	4 815 616	39 377	440 568
河　北 **Hebei**	268 506	841 551	8 173	15 963
唐　山 Tangshan	46 339	119 001	1 827	3 818
秦皇岛 Qinhuangdao	212 231	705 836	5 094	9 695
沧　州 Cangzhou	9 936	16 714	1 252	2 450
辽　宁 **Liaoning**	1 470 689	4 451 578	93 162	287 474
大　连 Dalian	947 705	3 209 687	46 287	160 706
丹　东 Dandong	264 863	614 898	1 065	2 395
锦　州 Jinzhou	105 520	283 605	20 769	57 430
营　口 Yingkou	59 187	129 591	329	737
盘　锦 Panjin	67 160	148 075	16 491	40 603
葫芦岛 Huludao	26 254	65 722	8 221	25 603
上　海 **Shanghai**	4 390 495	15 398 040	415 478	1 427 783
江　苏 **Jiangsu**	379 275	2 257 522	23 087	171 333
南　通 Nantong	267 716	1 520 366	12 506	71 506
连云港 Lianyungang	82 712	506 020	5 496	50 562
盐　城 Yancheng	28 847	231 136	5 085	49 265

3-20 续表1 continued

城　市 City	外国人 Foreigners		香港同胞 Hong Kong	
	人次数 Arrivals	人天数 Nights	人次数 Arrivals	人天数 Nights
浙　江 Zhejiang	3 101 651	8 132 037	631 374	1 608 382
杭　州 Hangzhou	1 572 838	4 381 824	278 358	898 027
宁　波 Ningbo	464 598	1 526 803	135 665	323 684
温　州 Wenzhou	254 612	644 368	30 588	65 168
嘉　兴 Jiaxing	357 762	627 156	78 939	115 966
绍　兴 Shaoxing	270 243	467 733	56 067	85 049
舟　山 Zhoushan	122 532	335 578	44 211	98 266
台　州 Taizhou	59 066	148 575	7 546	22 222
福　建 Fujian	903 089	5 326 623	735 266	3 905 791
福　州 Fuzhou	357 898	2 945 726	80 822	614 389
厦　门 Xiamen	430 759	2 037 116	135 183	541 310
莆　田 Putian	8 583	99 092	25 233	84 530
泉　州 Quanzhou	61 387	116 421	453 024	2 576 194
漳　州 Zhangzhou	42 718	124 789	40 749	88 978
宁　德 Ningde	1 744	3 479	255	390
山　东 Shandong	1 814 019	5 554 545	145 006	367 549
青　岛 Qingdao	801 424	2 323 891	105 950	253 790
东　营 Dongying	18 156	87 683	4 019	13 285
烟　台 Yantai	343 474	1 378 215	18 129	50 767
潍　坊 Weifang	139 230	549 910	13 974	43 563
威　海 Weihai	308 641	747 779	2 194	4 698
日　照 Rizhao	179 161	429 608	487	1 133
滨　州 Binzhou	23 933	37 459	253	313

3-20 续表2 continued

城 市 City	外国人 Foreigners		香港同胞 Hong Kong	
	人次数 Arrivals	人天数 Nights	人次数 Arrivals	人天数 Nights
广 东 **Guangdong**	5 884 089	13 782 089	14 747 801	29 019 740
广 州 Guangzhou	2 311 843	5 624 991	3 789 672	8 014 439
深 圳 Shenzhen	1 463 808	3 268 109	7 019 490	13 611 128
珠 海 Zhuhai	478 622	1 151 488	1 019 773	1 833 037
汕 头 Shantou	68 875	179 075	44 301	86 830
江 门 Jiangmen	150 002	382 697	522 493	1 011 256
湛 江 Zhanjiang	20 259	62 051	20 017	45 426
茂 名 Maoming	5 991	18 306	13 562	20 303
惠 州 Huizhou	338 653	844 777	919 953	1 310 346
汕 尾 Shanwei	442	679	25 860	27 502
阳 江 Yangjiang	5 454	13 635	28 373	59 583
东 莞 Dongguan	896 420	1 792 840	751 210	1 727 783
中 山 Zhongshan	109 589	363 835	244 867	539 687
潮 州 Chaozhou	30 685	73 644	313 424	671 369
揭 阳 Jieyang	3 446	5 962	34 806	61 051
广 西 **Guangxi**	87 596	119 094	30 647	47 335
北 海 Beihai	31 501	49 330	18 249	26 574
防城港 Fangchenggang	54 430	66 728	2 691	3 592
钦 州 Qinzhou	1 665	3 036	9 707	17 169
海 南 **Hainan**	294 723	1 030 494	89 655	181 758
海 口 Haikou	61 850	118 667	22 727	35 551
三 亚 Sanya	232 873	911 827	66 928	146 207

3-20 续表3 continued

城市 City	澳门同胞 Macau		台湾同胞 Taiwan Province	
	人次数 Arrivals	人天数 Nights	人次数 Arrivals	人天数 Nights
合　计 Total	**1 956 786**	**4 272 806**	**5 545 175**	**16 248 768**
天　津 Tianjin	6 255	28 015	58 840	427 379
河　北 Hebei	2 227	5 615	8 662	17 400
唐　山 Tangshan	898	2 375	1 350	3 379
秦皇岛 Qinhuangdao	559	1 225	6 322	12 225
沧　州 Cangzhou	770	2 015	990	1 796
辽　宁 Liaoning	29 567	85 230	95 380	303 116
大　连 Dalian	1 374	4 997	54 677	189 965
丹　东 Dandong	804	1 793	1 128	2 479
锦　州 Jinzhou	7 212	21 727	16 880	47 493
营　口 Yingkou	108	247	163	358
盘　锦 Panjin	14 649	38 208	16 700	43 376
葫芦岛 Huludao	5 420	18 258	5 832	19 445
上　海 Shanghai	17 816	65 392	510 146	2 191 104
江　苏 Jiangsu	3 960	12 079	48 558	328 843
南　通 Nantong	1 443	6 995	18 201	116 150
连云港 Lianyungang	2 459	4 960	9 409	94 076
盐　城 Yancheng	58	124	20 948	118 617

3-20 续表4 continued

城市 City	澳门同胞 Macau		台湾同胞 Taiwan Province	
	人次数 Arrivals	人天数 Nights	人次数 Arrivals	人天数 Nights
浙　江 **Zhejiang**	124 497	240 189	875 222	3 719 027
杭　州 Hangzhou	17 106	31 468	435 743	1 306 381
宁　波 Ningbo	41 927	108 525	158 358	443 652
温　州 Wenzhou	10 164	15 706	34 370	77 435
嘉　兴 Jiaxing	14 864	21 981	105 143	1 584 848
绍　兴 Shaoxing	35 328	52 818	70 075	112 742
舟　山 Zhoushan	3 137	6 817	53 602	122 614
台　州 Taizhou	1 971	2 874	17 931	71 355
福　建 **Fujian**	55 380	261 572	1 169 377	3 555 387
福　州 Fuzhou	4 273	23 819	186 321	462 859
厦　门 Xiamen	4 463	19 336	711 502	2 036 986
莆　田 Putian	16 065	58 540	105 503	140 728
泉　州 Quanzhou	27 233	152 052	85 077	383 824
漳　州 Zhangzhou	3 311	7 790	79 896	529 448
宁　德 Ningde	35	35	1 078	1 542
山　东 **Shandong**	34 713	88 816	143 043	410 994
青　岛 Qingdao	19 763	48 428	73 533	171 051
东　营 Dongying	1 305	4 822	1 923	6 853
烟　台 Yantai	9 827	24 909	39 471	153 604
潍　坊 Weifang	3 051	9 007	16 275	47 542
威　海 Weihai	550	1 176	11 291	30 866
日　照 Rizhao	203	453	294	746
滨　州 Binzhou	14	21	256	332

3-20 续表5 continued

城 市 City	澳门同胞 Macau		台湾同胞 Taiwan Province	
	人次数 Arrivals	人天数 Nights	人次数 Arrivals	人天数 Nights
广 东 **Guangdong**	1 674 303	3 471 480	2 587 084	5 202 661
广 州 Guangzhou	349 286	809 119	443 242	1 029 812
深 圳 Shenzhen	42 027	74 919	438 372	1 046 175
珠 海 Zhuhai	680 354	1 488 647	799 680	1 731 905
汕 头 Shantou	608	912	9 377	27 193
江 门 Jiangmen	414 420	777 427	72 024	162 132
湛 江 Zhanjiang	1 664	2 725	8 725	16 415
茂 名 Maoming	4 079	5 307	4 001	8 332
惠 州 Huizhou	29 012	38 519	154 905	258 830
汕 尾 Shanwei	7 477	7 951	2 862	4 397
阳 江 Yangjiang	4 212	8 424	5 629	14 073
东 莞 Dongguan	49 680	84 456	562 315	674 778
中 山 Zhongshan	74 771	144 308	46 542	157 871
潮 州 Chaozhou	16 312	27 998	29 243	52 930
揭 阳 Jieyang	401	768	10 167	17 818
广 西 **Guangxi**	6 092	9 670	14 133	23 080
北 海 Beihai	2 781	4 063	8 906	13 602
防城港 Fangchenggang	72	86	118	132
钦 州 Qinzhou	3 239	5 521	5 109	9 346
海 南 **Hainan**	1 976	4 748	34 730	69 777
海 口 Haikou	461	783	18 213	32 708
三 亚 Sanya	1 515	3 965	16 517	37 069

主要统计指标解释

1. 海洋捕捞产量 凡是从海洋里捕捞的天然生长的水产品产量为捕捞产量。

2. 海水养殖产量 凡是从人工投放苗种或天然纳苗并进行人工饲养管理的海水养殖水域中捕捞的水产品产量为海水养殖产量。

3. 远洋捕捞产量 由各远洋渔业企业和各生产单位按我国远洋渔业项目管理办法组织的远洋渔船（队）在非我国管辖海域（外国专属经济区水域或公海）捕捞的水产品产量。中外合资、合作渔船捕捞的水产品只统计按协议应属于中方所有的部分。

4. 原油产量 是按净原油量来计算的，能直接用于销售和生产自用的原油量。目前海洋石油系统原油产量计算方法采用倒算法。

原油产量=销售量+期末库存量-期初库存量+海上平台及陆地终端处理厂自用量。

5. 天然气产量 指进入集输管网的销售量和就地利用的全部气量。

天然气产量=外输（销）量+企业自用量

6. 海洋原油出口量 指销往国外的产品数量。

7. 海洋原油出口创汇额 指产品销往国外的归中方所有的全部外汇收入。以美元或万美元表示。

8. 造船综合吨 等于以计量单位载重吨和满载排水量吨的民用船舶的吨位数之和。

9. 货运量 指经船舶实际运送的货物重量，按到达量统计。

10. 货物周转量 指实际运送的货物与其运送距离的乘积。

11. 集装箱运量 既包含货重，也包含箱重。箱重系指承运租用的空箱重量凡有运费收入的空箱，其重量应统计为运量，按空箱 1 吨为货运量 1 吨计算；若无收入，所承运的空箱一律不作运量统计。

12. 旅客周转量 指实际运送的旅客人数与其运送距离的乘积。

13. 国际旅游外汇收入 入境旅游者在中国（大陆）境内旅行、游览过程中用于交通、参观游览、住宿、餐饮、购物、娱乐等的全部花费。

14. 接待人次数 指报告期内我国接待游客人数。游客按出游地分为入境游客（即海外游客）和国内游客，按出游时间分为旅游者（过夜游客）和一日游游客（不过夜游客）。

15. 接待人天数 指过夜旅游者的停留天数。

16. 外国人 指外国国籍的人，加入外籍的中国血统华人也计入外国人。

17. 港澳台同胞 指居住在我国香港特别行政区、澳门特别行政区和台湾省的中国同胞。

Explanatory Notes on Main Statistical Indicators

1. Marine Catches refers to the output of the naturally growing aquatic products caught from the sea.
2. Mariculture Production refers to the output of aquatic products whose young are artificially released or naturally collected, and raised and managed artificially, and which are caught from the waters of mariculture.
3. Deep-Sea Fishing Production refers to the output of aquatic products caught in the non-Chinese jurisdictional sea areas (foreign EEE or high sea) by the distant fishing vessels (fleet) organized by various distant fishing businesses and production units according to the management measures of the China distant fishing projects. The aquatic products caught by the Chinese-foreign joint ventures' and cooperative fishing vessels are counted only for the part owned by the Chinese side according to the agreement.
4. Output of Crude Oil is calculated on the basis of the net amount of crude oil, i.e., the amount of crude oil that may be directly used for sale and for the production itself.

Output of crude oil = Volume of sales + Reserves at the end of the period - Reserves at the beginning of the period +Amount for self-use on the platforms and in the terminal processing plants on land.

5. Output of Natural Gas refers to the total gas volume of the sales volume entering the oil collecting and transport pipeline network and that used locally.

Output of natural gas = Volume of sales or transport to other areas + Volume used by the enterprise itself

6. Export Volume of Offshore Crude Oil refers to the amount of products for sale abroad.
7. Foreign Exchange Earnings of Offshore Crude Oil refer to the total foreign exchange income from oil (gas) products for sale abroad which is owned by the Chinese side.
8. Comprehensive Tonnages of Shipbuilding refers to the sum of tonnage of civilian vessels with the deadweight capacity and full-load displacement as measured.
9. Freight Traffic refers to the weight of cargoes actually transported by vessels, which is counted according to the volume of arrival.
10. Cargoes Turnover Volume refers to the product of the actually transported cargoes and the transport distance.
11. Freight Volume of Containers includes the weight of both cargoes and container boxes. The weight of container boxes refers to the weight of empty containers rented for transport or having freight income and should be included in the freight volume, one ton of empty boxes equalling to one ton of freight volume. The empty boxes which have no income for transportation are not included in the freight volume.

12. Passenger Turnover Volume refers to the product of the number of passengers actually transported and the shipping distance.

13. International Tourism (Foreign Exchange) Receipts refer to the total expenditure made by inbound tourists within the territory of China (the mainland) in their course of travel on transport, tours and sightseeing, lodging, food and beverage, shopping, entertainment, etc.

14. Number of Person-Times Received refers to the number of tourists received by China in the period reported. Visitors are divided into foreigner tourists and domestic visitors by origin of the travel, and tourists (overnight tourists) and same-day tourists (non-overnight tourists) by their length of stay.

15. Number of the Days of Stay refers to the number of the days of stay of overnight tourists.

16. Foreigners refer to the people with foreign nationality, including foreign nationals of Chinese descent.

17. Compatriots from Hong Kong, Macau and Taiwan Province refer to the Chinese compatriots living in the Hong Kong Special Administrative Region, the Macau Special Administrative Region and Taiwan Province.

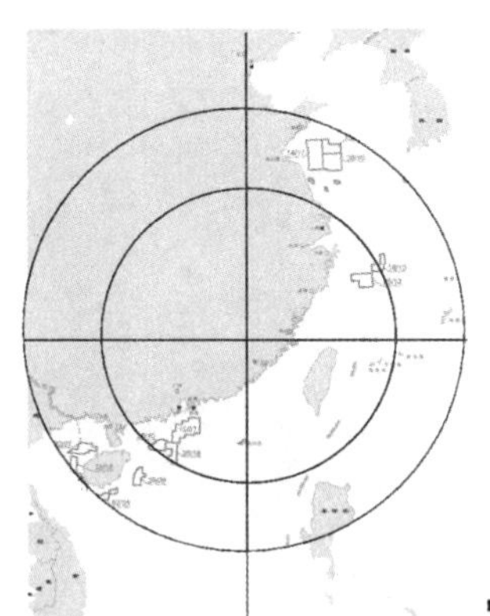

4

主要海洋产业生产能力
Production Capacity of Major Marine Industries

4-1 浅海滩涂海湾可养殖面积
Cultiratable Area of Shellow Sea, Tidal Flat and Bay

单位：千公顷 (1 000 hm²)

地区 Region		海水可养殖面积 Cultivatable Marine Area	浅海 Shallow Sea	滩涂 Tidal Flat	港湾 Bay
合计	**Total**	**2 599.67**	**1 622.55**	**796.56**	**180.55**
天津	Tianjin	18.49	10.00	8.49	
河北	Hebei	111.37	49.66	61.70	
辽宁	Liaoning	725.84	590.44	92.45	42.95
上海	Shanghai	3.22		3.22	
江苏	Jiangsu	139.00	7.87	130.96	0.17
浙江	Zhejiang	101.46	36.30	57.39	7.77
福建	Fujian	184.94	77.39	100.76	6.79
山东	Shandong	358.21	131.68	173.41	53.12
广东	Guangdong	835.67	664.00	120.00	51.67
广西	Guangxi	31.95	6.78	22.09	3.08
海南	Hainan	89.52	48.43	26.09	15.00

4-2 沿海地区海水养殖面积
Mariculture Area by Coastal Regions

单位：公顷 (hm²)

地　区 Region	海水养殖面积 Mariculture Area
合　计 Total	**1 865 944**
天　津 Tianjin	4 304
河　北 Hebei	121 013
辽　宁 Liaoning	630 700
上　海 Shanghai	
江　苏 Jiangsu	172 754
浙　江 Zhejiang	94 514
福　建 Fujian	133 942
山　东 Shandong	441 403
广　东 Guangdong	194 766
广　西 Guangxi	57 300
海　南 Hainan	15 248

4-3 海洋油田生产井情况
Survey of Offshore Oilfield Production Wells

单位：口 (number)

地　区 Region	合　计 Total	采油井 Oil Wells	采气井 Gas Wells	注水井 Injection Wells	其他井 Others
合　计　Total	**4 215**	**3 193**	**204**	**760**	**58**
天　津　Tianjin	2 413	1 795	101	472	45
河　北　Hebei	546	428		118	
辽　宁　Liaoning	264	203	13	35	13
上　海　Shanghai	30	12	18		
山　东　Shandong	476	352	8	116	
广　东　Guangdong	486	403	64	19	

4-4 海洋石油勘探情况
Work Volume of Offshore Oil Exploration

地　区 Region	地震测线 Seismic Line		钻井（口） Drilling (Well)	
	二维（公里） Two Dimensions (km)	三维（平方公里） Three Dimensions (km^2)	预探井 Wildcat Wells	评价井 Appraisal Wells
合　计 Total	**30 306**	**12 279**	**97**	**76**
天　津 Tianjin	0	5 477	40	29
其中：合作 Including: Cooperative	0	0	2	0
河　北 Hebei	0	120	19	15
辽　宁 Liaoning	0	164	4	3
上　海 Shanghai	8 803	1 853	1	0
其中：合作 Including: Cooperative	0	1 853	0	0
山　东 Shandong	394	321	5	9
广　东 Guangdong	21109	4344	28	20
其中：合作 Including: Cooperative	1900	986	7	5

4-5 沿海地区盐田面积和海盐生产能力
Salt Pan Area and Sea Salt Production Capacity by Coastal Regions

地 区 Region	盐田总面积（公顷） Total Area of Salt Pan (hm^2)		生产面积（公顷） Production Area (hm^2)		年末海盐生产能力（万吨） Year-End Capacity of Sea Salt Production (10 000 tons)	
	2008	2009	2008	2009	2008	2009
合 计 Total	**470 681**	**465 974**	**336 793**	**324 453**	**3 816.53**	**4 194.04**
天 津 Tianjin	31 977	32 501	30 788	30 866	188.69	196.39
河 北 Hebei	81 745	78 613	77 012	67 871	520.86	527.46
辽 宁 Liaoning	50 030	45 146	44 784	39 340	268.37	300.48
江 苏 Jiangsu	72 816	70 218	16 437	24 729	133.00	126.00
浙 江 Zhejiang	3 789	3 643	3 172	2 988	15.76	16.96
福 建 Fujian	6 921	6 029	5 742	5 501	49.58	50.00
山 东 Shandong	205 702	212 006	154 776	142 170	2 588.10	2 927.55
广 东 Guangdong	10 484	10 579	6 506	6 368	19.48	16.11
广 西 Guangxi	3 460	3 552	1 260	1 778	13.99	14.59
海 南 Hainan	3 757	3 687	2 822	2 842	18.70	18.50

4-6 沿海地区风能发电能力（2008年）
Wind Power Production by Coastal Regions, 2008

单位：万千瓦　　(10 000 kW)

地　区 Region	风能年发电能力 Annual Wind Power Generation Capacity
合　计 Total	**239.06**
辽　宁 Liaoning	34.00
河　北 Hebei	11.25
上　海 Shanghai	3.94
江　苏 Jiangsu	43.78
浙　江 Zhejiang	19.06
福　建 Fujian	28.38
山　东 Shandong	56.14
广　东 Guangdong	36.69
海　南 Hainan	5.82

4-7 主要潮汐电站分布情况
Distribution of Major Tidal Power Stations

电站名称 Name	运行情况 Status of operation	装机容量（千瓦） Capacity (kW)
江厦潮汐试验电站	1970年开始建造，1980年投入使用，运行至今。	3 900
Jiangxia Experimental Tidal Power Station	It began construction in 1970, was put into use in 1980, and has been in operation so far.	
海山潮汐电站	1972年开始建造，1975年投入使用，运行至今。	250
Haishan Tidal Power Station	It began construction in 1972, was put into use in 1975, and has been in operation so far.	
岳浦潮汐电站	1970年开始建造，1978年停止运行。	
Yuepu Tidal Power Station	It began construction in 1970, stopped power generation in 1978.	
白沙口潮汐电站	1970年开始建造，1978年投入使用，运行至今。	960
Baishakou Tidal Power Station	It began construction in 1970, was put into use in 1978, and has been in operation so far.	
果子山潮汐电站	已经投入前期筹备工作阶段，未确定开工时间。	
Guozishan Tidal Power Station	It has been put into the first stage of preparation, and the openning date has not been fixed.	

4-8 主要海上活动船舶
Major Vessels Operating on the Sea

类　别 Type	艘数 （艘） Number of Vessels (unit)	总吨 （万吨位） grt (10 000 tons)	净载重量 （万吨） Net Weight Tonnage (10 000 tons)	载客量 （客位） Passenger Spaces (seat)	总功率 （千瓦） Total Power (kW)
一 、海洋生产用船 Vessels for Marine Production					
海洋渔业船舶 Marine Fishing Vessels	295 847	684. 35			15 493 687
远洋渔船 Ocean-going Fishing Vessels	1 456				861 328
海洋油气船舶 Offshore Oil and Gas Vessels	172	132. 60	83. 20	8 275	997 781
钻井平台 Drilling Vessels	25	21. 00	6. 00	2 858	142 434
物探船 Physical Exploration Vessels	8	0. 60	0. 20	246	9 350
其 他 Others	139	111. 00	77. 00	5 171	845 997
海洋运输船舶 Maritime Transport Vessels	12 097	58 353 752	86 202 137	176 400	25 356 011
二、海洋科研用船 Vessels for Marine Scientific Research					
海洋地质勘探船 Marine Geology Survey Vessels	6	1. 18	0. 37	219	22 588
海洋调查船 Marine Research Vessels	10	3. 62	1. 67	562	46 931
中国科学院 Chinese Acadamy of Sciences	6	1. 02	0. 32	244	21 919
国家海洋局 State Oceanic Administration	4	2. 60	1. 35	318	25 012
三、海洋执法用船 Vessels for Marine Law Enforcement					
#海监船 Marine Monitoring Vessels	44	3. 50			73 297

4-9 沿海规模以上港口生产用码头泊位
Berths for Productive Use at above Designed Size Seaports

单位：米, 个 (m, number)

港口 Seaport		码头长度 Length of Quay Line	泊位个数 Number of Berths	#万吨级 10 000 Tonnage Class
合 计	**Total**	**560 565**	**4 516**	**1 214**
丹 东	Dandong	3 677	27	12
大 连	Dalian	33 318	196	78
营 口	Yingkou	11 187	50	34
锦 州	Jinzhou	4 698	18	16
秦皇岛	Qinhuangdao	14 750	66	42
黄 骅	Huanghua	3 188	16	10
唐 山	Tangshan	10 901	44	41
天 津	Tianjin	26 736	124	80
烟 台	Yantai	13 866	74	45
威 海	Weihai	1 554	9	4
青 岛	Qingdao	18 749	73	57
日 照	Rizhao	10 914	44	39
连云港	Lianyungang	10 505	53	40
上 海	Shanghai	72 274	614	148
宁波-舟山	Ningbo-Zhoushan	64 630	628	108
嘉 兴	Jiaxing	5 226	38	16
台 州	Taizhou	10 120	168	4
温 州	Wenzhou	15 987	232	15
宁 德	ningde	4 226	42	2
福 州	Fuzhou	15 554	121	40

4-9 续表 continued

港 口 Seaport		码头长度 Length of Quay Line	泊位个数 Number of Berths	#万吨级 10 000 Tonnage Class
厦 门	Xiamen	18 266	101	50
漳 州	Zhangzhou	1 514	17	1
莆 田	Putian	3 881	42	4
泉 州	Quanzhou	12 765	99	17
汕 头	Shantou	8 752	82	17
汕 尾	Shanwei	1 580	14	1
惠 州	Huizhou	6 200	35	16
深 圳	Shenzhen	28 937	156	66
虎 门	humen	9 790	87	15
广 州	Guangzhou	41 042	467	58
中 山	Zhongshan	3 447	60	0
珠 海	Zhuhai	11 600	118	15
江 门	Jiangmen	9 196	144	2
阳 江	Yangjiang	1 407	7	6
茂 名	Maoming	2 333	18	8
湛 江	Zhanjiang	14 152	148	30
北部湾	Beibuwan	23 202	211	46
洋 浦	Yangpu	4 460	23	14
海 口	Haikou	4 422	41	10
八 所	Basuo	1 559	9	7

4-10 主要沿海城市星级饭店基本情况
Star-rated Hotels and Occupancies in Major Coastal Cities

城　市 City	饭店数（座） Number of Hotels (unit)	客房数（间） Number of Rooms (unit)	床位数（张） Number of Beds (unit)	客房出租率（%） Room Occupancy (%)
合　计　Total	**3 750**	**518 534**	**906 386**	**56.45**
天　津 **Tianjin**	111	17 682	29 507	46.37
河　北 **Hebei**	166	21 528	40 239	56.22
唐　山 Tangshan	61	7 006	13 526	54.66
秦皇岛 Qinhuangdao	70	10 113	18 891	49.28
沧　州 Cangzhou	35	4 409	7 822	76.18
辽　宁 **Liaoning**	296	32 195	57 792	61.50
大　连 Dalian	177	21 328	36 899	57.81
丹　东 Dandong	46	3 968	7 929	75.71
锦　州 Jinzhou	19	1 781	3 483	76.00
营　口 Yingkou	17	1 585	2 899	69.95
盘　锦 Panjin	13	1 364	2 334	69.17
葫芦岛 Huludao	24	2 169	4 248	63.95
上　海 **Shanghai**	298	61 259	96 937	50.24
江　苏 **Jiangsu**	202	17 624	32 519	60.44
南　通 Nantong	68	7 198	13 847	61.83
连云港 Lianyungang	77	5 781	10 393	56.27
盐　城 Yancheng	57	4 645	8 279	62.77
浙　江 **Zhejiang**	855	112 871	202 607	60.23
杭　州 Hangzhou	307	48 951	93 861	62.26
宁　波 Ningbo	191	22 964	37 648	57.41
温　州 Wenzhou	92	11 264	18 933	66.80
嘉　兴 Jiaxing	57	7 459	12 403	54.91
绍　兴 Shaoxing	86	10 169	18 324	57.16
舟　山 Zhoushan	59	4 535	8 499	58.29
台　州 Taizhou	63	7 529	12 939	56.97
福　建 **Fujian**	291	38 183	66 258	67.95
福　州 Fuzhou	74	11 169	18 702	71.90
厦　门 Xiamen	76	12 226	20 302	61.21

4-10 续表 continued

城 市 City	饭店数（座） Number of Hotels (unit)	客房数（间） Number of Rooms (unit)	床位数（张） Number of Beds (unit)	客房出租率（%） Room Occupancy (%)
莆 田 Putian	10	1 376	2 452	53.36
泉 州 Quanzhou	87	9 578	18 019	78.45
漳 州 Zhangzhou	16	1 460	2 578	49.90
宁 德 Ningde	28	2 374	4 205	93.54
山 东 **Shandong**	525	66 196	121 904	52.33
青 岛 Qingdao	193	27 250	48 611	45.37
东 营 Dongying	24	3 116	5 485	65.36
烟 台 Yantai	99	12 988	24 239	66.99
潍 坊 Weifang	54	5 821	10 721	78.46
威 海 Weihai	94	9 568	18 503	52.42
日 照 Rizhao	42	5 385	10 511	57.97
滨 州 Binzhou	19	2 068	3 834	61.07
广 东 **Guangdong**	788	117 689	194 205	56.06
广 州 Guangzhou	207	36 947	65 838	57.62
深 圳 Shenzhen	153	25 445	38 810	56.50
珠 海 Zhuhai	89	11 508	19 230	0.00
汕 头 Shantou	41	6 694	11 318	56.72
江 门 Jiangmen	28	3 120	5 326	56.86
湛 江 Zhanjiang	45	4 019	7 361	58.02
茂 名 Maoming	16	1 606	2 836	51.31
惠 州 Huizhou	60	5 939	9 151	52.31
汕 尾 Shanwei	14	1 691	3 436	67.07
阳 江 Yangjiang	28	1 443	2 707	53.22
东 莞 Dongguan	62	13 234	18 185	49.85
中 山 Zhongshan	31	4 026	6 530	40.26
潮 州 Chaozhou	5	494	878	71.75
揭 阳 Jieyang	9	1 523	2 599	62.89
广 西 **Guangxi**	70	7 872	14 439	59.62
北 海 Beihai	38	4 467	8 321	57.68
防城港 Fangchenggang	14	1 298	2 356	66.17
钦 州 Qinzhou	18	2 107	3 762	61.01
海 南 **Hainan**	148	25 435	49 979	63.18
海 口 Haikou	72	11 682	20 671	58.89
三 亚 Sanya	76	13 753	29 308	66.74

4-11 沿海地区旅行社数

Number of Travel Agencies by Coastal Regions

单位：家 (number)

地　区 Region	旅行社总数 Number of Travel Agencies
合　计　Total	**10 290**
天　津　Tianjin	267
河　北　Hebei	1 060
辽　宁　Liaoning	1 001
上　海　Shanghai	851
江　苏　Jiangsu	1 646
浙　江　Zhejiang	1 447
福　建　Fujian	626
山　东　Shandong	1 752
广　东　Guangdong	1 047
广　西　Guangxi	385
海　南　Hainan	208

主要统计指标解释

1. 海水可养殖面积 指利用滩涂、浅海、港湾进行鱼、虾、蟹、贝、藻等海水经济动植物的人工养殖的水面面积。

2. 浅海养殖 指在可养殖的浅海中进行海水经济动植物养殖。

3. 港湾养殖 指利用港、湾，或在海边、河口附近的滩涂、洼地拦闸筑堤进行海水养殖。

4. 海水养殖面积 是指利用海上、滩涂、陆基进行鱼、甲壳类（虾、蟹）、贝、藻等海水经济动植物的人工养殖的水面面积。在报告期内无论是否全部收获或尚未收获其产品，均应统计在海水养殖面积中。但有些滩涂、水面不投放苗种或投放少量苗种，只进行一般管理的，不统计为养殖面积。

5. 盐田总面积 指盐田占有的全部面积。包括储卤、蒸发、保卤、结晶面积、滩内的沟、壕、池、埝、滩坨等面积及滩外的沟、壕、公路及杂地面积。

6. 生产面积 指直接提供给海盐生产的面积，包括结晶面积、蒸发面积、保卤面积，滩内的沟、壕、池、埝面积及滩坨面积。

7. 年末海盐生产能力 指年末企业生产原盐的全部设备的综合平衡能力。海盐生产露天作业，受天气影响，因而计算生产能力时，成熟滩田按 10 年实际平均单位生产面积产量乘以本年成熟滩田生产面积而得，新滩田按设计能力及滩田成熟程度可能达到的产量计算。

8. 海洋渔业船舶 是指配置机器作为动力的从事海洋渔业生产和辅助渔业生产的船舶。

9. 远洋渔船 按我国远洋渔业项目管理办法在非我国管辖海域（外国专属经济区水域或公海）进行常年或季节性生产的渔船。

10. 泊位个数 是指设有系靠船舶设施，在同一时间内可供靠泊最大吨级船舶的艘数。即可靠泊一艘船舶，则计为一个泊位，余类推。泊位分码头泊位和浮筒泊位。

11. 客房数 指饭店实际可用于接待旅游者的房间数。

12. 床位数 指饭店实际可用于接待旅游者的床位数。

Explanatory Notes on Main Statistical Indicators

1. Marine Cultivatable Area refers to water areas in tidal flat, shallow sea and bay that is used to breed marine economic animals and plants, such as fish, shrimp, crab, shellfish, alga and so on.

2. Shallow Sea Cultivation refers to the breeding of marine economic animals and plants in the cultivatable shallow sea.

3. Harbor Cultivation refers to the marine cultivation conducted in harbors, bays, or the tidal flat or marshes around seaside and bayou by blocking the gate and banking up the dam.

4. Mariculture Area refers to the area of the water surface where seawater economic animals and plants, such as crustacean (shrimp, crab), shellfish and algae, are cultivated at sea, on tidal flat and land. Whether or not all the products in the area have been harvested or the products have not been harvested yet in the period covered by the report, the area is included in the Mariculture Area. But some tidal flats and water surfaces where none or a small amount of the young have been released and only general management is carried out are not included in the Mariculture Area.

5. Total Area of Salt Pans refers to the total area covered by salt pans, including the area for brine storage, evaporation, brine preservation, and crystallization, the area of ditches, moats, pondsand banks within the beach as well as beach mounds, and ditches, moats, highway beyond the beach as well as the area of miscellaneous lands.

6. Area of Salt Pan Production refers to the area directly provided for sea-salt productions, including the area for crystallization, evaporation and brine preservation, the area of ditches, moats, ponds and banks within the beach as well as the area of beach mounds.

7. Year-End Capacity of Crude Salt Production refers to the integrated and balanced capacity of all equipment of the enterprise used for crude salt production at the end of the year. As sea salt production is an open-air operation, which is subject to the effect of weather, the production capacity of a matured salt pan is calculated at the productions of the actual average unit production area in ten years times the production area of the matured salt pan in the current year. The production capacity of new salt pans is calculated at the production that may be reached in the light of the designed capacity and the level of maturity of the salt pan.

8. Marine Fishing Vessels refer to the vessels equipped with machines as motive power and going for marine fishery production and auxiliary fishery production.

9. Deep Sea Fishing Vessels refer to the fishing vessels which carry out production all the year round or seasonally in the non-Chinese jurisdictional, sea areas (foreign EEZ or high sea) according to the China Deep-Sea Fishing Projects Management Measures.

10. Number of Berths refers to the spaces equipped with facilities for docking ships and the number of ships of the maximum tonnage that may dock or anchor in them. A space for a ship to dock is counted as one berth and the rest are reasoned out by analogy. Berths are divided into wharf berths and buoy berths.

11. Number of Rooms refers to the number of guest rooms actually used by the hotels receiving tourists.

12. Number of Beds refers to the number of beds actually used by the hotels receiving tourists.

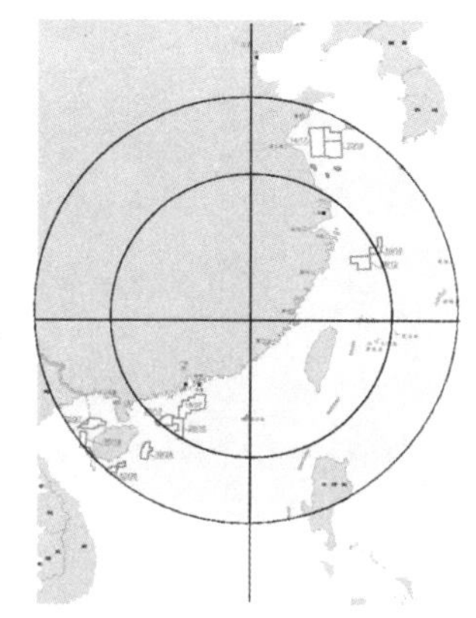

5

涉海就业

Ocean-Related Employment

5-1 沿海地区涉海就业人员情况
Ocean-Related Employed Personnel by Coastal Regions

地 区 Region	2001		2008		2009	
	总数（万人） Total Number (10 000 persons)	占地区就业人员比重（%） Proportion in the Regional Employed Personnel (%)	总数（万人） Total Number (10 000 persons)	占地区就业人员比重（%） Proportion in the Regional Employed Personnel (%)	总数（万人） Total Number (10 000 persons)	占地区就业人员比重（%） Proportion in the Regional Employed Personnel (%)
合 计 Total	**2 107.6**	**8.1**	**3 218.3**	**10.3**	**3 270.6**	**10.1**
天 津 Tianjin	106.4	25.9	162.5	32.3	165.1	32.6
河 北 Hebei	58.0	1.7	88.6	2.4	90.0	2.3
辽 宁 Liaoning	196.0	10.7	299.3	14.3	304.2	13.9
上 海 Shanghai	127.5	18.4	194.7	21.7	197.9	21.3
江 苏 Jiangsu	116.9	3.3	178.5	4.1	181.4	4.0
浙 江 Zhejiang	256.4	9.2	391.5	10.6	397.9	10.4
福 建 Fujian	259.7	15.5	396.6	19.1	403.0	18.6
山 东 Shandong	319.9	6.8	488.5	9.1	496.4	9.1
广 东 Guangdong	505.3	12.8	771.6	14.1	784.1	13.9
广 西 Guangxi	68.9	2.7	105.2	3.7	106.9	3.7
海 南 Hainan	80.6	23.7	123.1	29.9	125.1	29.0
其 他* Others	12.0		18.3		18.6	

注：2008年和2009年为推算数据；其他为非沿海地区涉海就业人员数。

Note:The data for 2008 and 2009 are the calculated ones; Others refer to the number of ocean-related employed persons in the non-coastal regions.

5-2 全国主要海洋产业就业人员情况
Employed Personnel in the Major Marine Industries Throughout the Country

单位：万人 (10 000 persons)

海洋产业 Marine Industry	2001	2008	2009
合计 Total	**719.1**	**1 097.0**	**1 115.0**
海洋渔业及相关产业 Marine Fishery and the Related Industries	348.3	531.3	540.0
海洋石油和天然气业 Offshore Oil and Natural Gas Industry	12.4	18.9	19.2
海滨砂矿业 Beach Placer Industry	1.0	1.5	1.6
海洋盐业 Sea Salt Industry	15.0	22.9	23.3
海洋化工业 Marine Chemical Industry	16.1	24.6	25.0
海洋生物医药业 Marine Biomedicine	0.6	0.9	0.9
海洋电力和海水利用业 Marine Electric Power and Seawater Utilization Industry	0.7	1.1	1.1
海洋船舶工业 Marine Shipbuilding Industry	20.6	31.4	31.9
海洋工程建筑业 Marine Engineering Architecture	38.8	59.2	60.2
海洋交通运输业 Maritime Communications and Transportation Industry	50.8	77.5	78.8
滨海旅游业 Coastal Tourism	78.3	119.5	121.4
其他海洋产业 Other Marine Industries	136.5	208.2	211.6

注：2008年和2009年为推算数据。

Note: The data for 2008 and 2009 are the caculated ones.

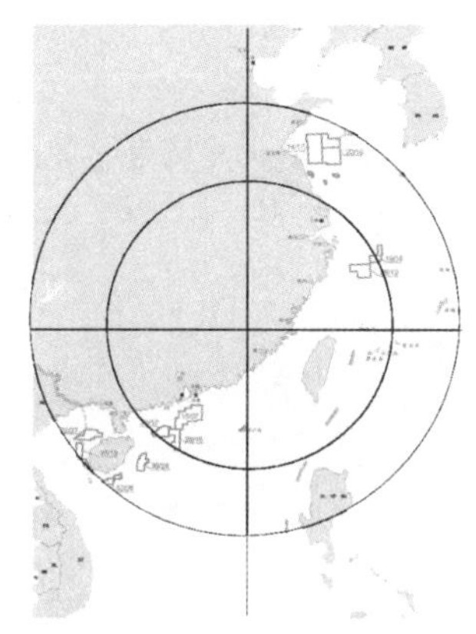

6

海洋科学技术

Marine Science and Technology

6-1 分行业海洋科研机构及人员情况
Marine Scientific Research Institutions and Personnel by Industry

行　业 Industry	机构数（个） Number of Institutions(unit)	从业人员（人） Employed Population(person)
合　计 Total	**186**	**34 076**
海洋基础科学研究 Marine Basic Scientific Research	**105**	**15 961**
海洋自然科学 Marine Natural Science	56	11 573
海洋社会科学 Marine Social Science	5	1 087
海洋农业科学 Marine Agricultural Science	40	3 201
海洋生物医药 Marine Biomedicine	4	100
海洋工程技术研究 Marine Engineering Technology Research	**68**	**16 294**
海洋化学工程技术 Marine Chemical Engineeing Technology	13	5 464
海洋生物工程技术 Marine Bioengineering Technology	2	209
海洋交通运输工程技术 Marine Communications and Transport Technology	16	3 339

6-1 续表 continued

行 业 Industry	机构数（个） Number of Institutions(unit)	从业人员（人） Employed Population(person)
海洋能源开发技术 Marine Energies Development Technology	4	2 594
海洋环境工程技术 Marine Environmental Engineering Technology	11	877
河口水利工程技术 Eustuarine Water Conservancy Engineering Technology	18	2 934
其他海洋工程技术 Other Marine Engineering Technology	4	877
海洋信息服务业 **Marine Information Service**	**10**	**1 130**
其他海洋信息服务 Other Marine Information Services	10	1 130
海洋技术服务业 **Marine Technological Service Industry**	**3**	**691**
其他海洋专业技术服务 Other Marine Professional and Technological Services	2	119
海洋工程管理服务 Marine Engineering Management Service	1	572

注：机构为县级以上科研机构；行业分类参照《海洋及相关产业分类》标准。

Note: The institutions are the scientific research institutions above the county level;The classification of industries follows the standard *Classification of Marine Industries and the Related Industries*.

6-2 分行业海洋科研机构科技活动人员学历构成

Educational Background Composition of the Personnel Engaged in Scientific and Technological Activities in Marine Scientific Research Institutions by Industry

单位：人 (person)

行 业 Industry	科技活动人员 Personnel Engaged in Scientific Activities	研究生 Postgraduate	#博士 Doctor	大学生 Graduate	大专生 College Graduates	其他 Others
合 计 Total	**27 888**	**11 453**	**4 599**	**9 866**	**4 200**	**2 369**
海洋基础科学研究 Marine Basic Scientific Research	**13 419**	**6 072**	**3 023**	**4 362**	**1 713**	**1 272**
海洋自然科学 Marine Natural Science	9 848	4 958	2 714	2 914	1 114	862
海洋社会科学 Marine Social Science	1 011	408	88	375	173	55
海洋农业科学 Marine Agricultural Science	2 477	694	221	1 021	411	351
海洋生物医药 Marine Biomedicine	83	12	0	52	15	4
海洋工程技术研究 Marine Engineering Technology Research	**12 927**	**4 882**	**1 499**	**4 716**	**2 324**	**1 005**
海洋化学工程技术 Marine Chemical Engineeing Technology	4 610	1 363	475	1 545	1 079	623
海洋生物工程技术 Marine Bioengineering Technology	181	61	6	74	28	18

6-2 续表 continued

行 业 Industry	科技活动人员 Personnel Engaged in Scientific Activities	研究生 Postgraduate	#博士 Doctor	大学生 Graduate	大专生 College Graduates	其他 Others
海洋交通运输工程技术 Marine Communications and Transport Technology	2 294	674	68	951	560	109
海洋能源开发技术 Marine Energies Development Technology	1 902	1 144	540	472	187	99
海洋环境工程技术 Marine Environmental Engineering Technology	786	251	37	432	67	36
河口水利工程技术 Eustuarine Water Conservancy Engineering Technology	2 316	959	323	957	306	94
其他海洋工程技术 Other Marine Engineering Technology	838	430	50	285	97	26
海洋信息服务业 Marine Information Service	**876**	**257**	**51**	**422**	**117**	**80**
其他海洋信息服务 Other Marine Information Services	876	257	51	422	117	80
海洋技术服务业 Marine Technological Service Industry	**666**	**242**	**26**	**366**	**46**	**12**
其他海洋专业技术服务 Other Marine Professional and Technological Services	103	13	2	68	13	9
海洋工程管理服务 Marine Engineering Management Service	563	229	24	298	33	3

6-3 分行业海洋科研机构科技活动人员职称构成
Technical Title Composition of Personnel Engaged in Scientific and Technological Activities in the Marine Scientific Research Institutions by Industry

单位：人 (person)

行业 Industry	科技活动人员 Personnel Engaged in Scientific Activities	高级职称 Senior	中级职称 Intermediate	其他 Others
合计 Total	**27 888**	**10 398**	**8 694**	**8 796**
海洋基础科学研究 Marine Basic Scientific Research	**13 419**	**5 347**	**4 162**	**3 910**
海洋自然科学 Marine Natural Science	9 848	4 203	2 965	2 680
海洋社会科学 Marine Social Science	1 011	333	328	350
海洋农业科学 Marine Agricultural Science	2 477	792	849	836
海洋生物医药 Marine Biomedicine	83	19	20	44
海洋工程技术研究 Marine Engineering Technology Research	**12 927**	**4 540**	**3 990**	**4 397**
海洋化学工程技术 Marine Chemical Engineeing Technology	4 610	1 529	1 396	1 685
海洋生物工程技术 Marine Bioengineering Technology	181	46	75	60

6-3 续表 continued

行 业 Industry	科技活动人员 Personnel Engaged in Scientific Activities	高级职称 Senior	中级职称 Intermediate	其他 Others
海洋交通运输工程技术 Marine Communications and Transport Technology	2 294	576	594	1 124
海洋能源开发技术 Marine Energies Development Technology	1 902	951	504	447
海洋环境工程技术 Marine Environmental Engineering Technology	786	233	334	219
河口水利工程技术 Eustuarine Water Conservancy Engineering Technology	2 316	960	745	611
其他海洋工程技术 Other Marine Engineering Technology	838	245	342	251
海洋信息服务业 Marine Information Service	**876**	**293**	**325**	**258**
其他海洋信息服务 Other Marine Information Services	876	293	325	258
海洋技术服务业 Marine Technological Service Industry	**666**	**218**	**217**	**231**
其他海洋专业技术服务 Other Marine Professional and Technological Services	103	46	30	27
海洋工程管理服务 Marine Engineering Management Service	563	172	187	204

6-4 分行业海洋科研机构经费收入
Routine Fund Receipts of the Marine Scientific Research Institutions by Industry

单位：万元 (10 000 yuan)

行 业 Industry	经费收入总额 Fund Total	经常费 Routine Fund	科技活动借贷款 Loans in the scientific and technological activities	基本建设中政府投资 Government investment in the capital construction
合 计 **Total**	**1 601 610**	**1 487 639**	**1 334**	**112 637**
海洋基础科学研究 **Marine Basic Scientific Research**	**801 520**	**720 210**	**34**	**81 277**
海洋自然科学 Marine Natural Science	682 441	610 844	34	71 564
海洋社会科学 Marine Social Science	31 818	30 166	0	1 652
海洋农业科学 Marine Agricultural Science	85 951	77 890	0	8 062
海洋生物医药 Marine Biomedicine	1 310	1 310	0	0
海洋工程技术研究 **Marine Technological Service Industry**	**732 355**	**700 674**	**1 300**	**30 382**
海洋化学工程技术 Marine Chemical Engineeing Technology	238 671	236 878	1 300	493
海洋生物工程技术 Marine Bioengineering Technology	13 627	13 627	0	0

6-4 续表 continued

行 业 Industry	经费收入总额 Fund Total	经常费 Routine Fund	科技活动借贷款 Loans in the scientific and technological activities	基本建设中政府投资 Government investment in the capital construction
海洋交通运输工程技术 Marine Communications and Transport Technology	111 137	99 952	0	11 186
海洋能源开发技术 Marine Energies Development Technology	1 937 220	192 637	0	1 085
海洋环境工程技术 Marine Environmental Engineering Technology	37 770	37 757	0	13
河口水利工程技术 Eustuarine Water Conservancy Engineering Technology	109 268	91 662	0	17 606
其他海洋工程技术 Other Marine Engineering Technology	28 161	28 161	0	0
海洋信息服务业 **Marine Information Service**	**40 873**	**39 925**	**0**	**948**
其他海洋信息服务 Other Marine Information Services	40 873	39 925	0	948
海洋技术服务业 **Marine Technological Service Industry**	**26 862**	**26 831**	**0**	**30**
其他海洋专业技术服务 Other Marine Professional and Technological Services	5 099	5 068	0	30
海洋工程管理服务 Marine Engineering Management Service	21 763	21 763	0	0

6-5 分行业海洋科研机构科技课题情况
Marine Science and Technology Research Projects of the Research Institutions by Industry

单位：项 (item)

行　业 Industry	课题数 Number of research subjectss	基础研究 Basic Research	应用研究 Applied Research	试验发展 Experimental Development	成果应用 Result Application	科技服务 Scientific and Technological Service
合 计 Total	**12 600**	**2 753**	**3 338**	**2 550**	**1 074**	**2 885**
海洋基础科学研究 Marine Basic Scientific Research	**8 940**	**2 706**	**2 990**	**1 387**	**590**	**1 267**
海洋自然科学 Marine Natural Science	7 388	2 587	2 607	941	338	915
海洋社会科学 Marine Social Science	244	2	5	98	8	131
海洋农业科学 Marine Agricultural Science	1 274	116	373	335	230	220
海洋生物医药 Marine Biomedicine	34	1	5	13	14	1
海洋工程技术研究 Marine Engineering Technology Research	**3 479**	**47**	**343**	**1 141**	**479**	**1 469**
海洋化学工程技术 Marine Chemical Engineering Technology	531	6	50	322	120	33
海洋生物工程技术 Marine Bioengineering Technology	28	0	0	10	9	9

6-5 续表 continued

行 业 Industry	课题数 Number of research subjectss	基础研究 Basic Research	应用研究 Applied Research	试验发展 Experimental Development	成果应用 Result Application	科技服务 Scientific and Technologica l Service
海洋交通运输工程技术 Marine Communications and Transport Technology	640	3	30	134	142	331
海洋能源开发技术 Marine Energies Development Technology	611	18	113	305	31	144
海洋环境工程技术 Marine Environmental Engineering Technology	219	0	25	55	27	112
河口水利工程技术 Eustuarine Water Conservancy Engineering	1 396	18	117	300	148	813
其他海洋工程技术 Other Marine Engineering Technology	54	2	8	15	2	27
海洋信息服务业 Marine Information Service	**91**	**0**	**0**	**6**	**1**	**84**
其他海洋信息服务 Other marine information Services	91	0	0	6	1	84
海洋技术服务业 Marine Technological Service Industry	**90**	**0**	**5**	**16**	**4**	**65**
其他海洋专业技术服务 Other Marine Professional and Technological Services	12	0	4	5	3	0
海洋工程管理服务 Marine Engineering Management Service	78	0	1	11	1	65

6-6 分行业海洋科研机构科技论著情况
Marine Scientific and Technological Works of the Research Institutions by Industry

行　业 Industry	发表科技论文（篇） Scientific Theses Published (piece)	#国外发表 Published Abroad	出版科技著作（种） Scientific and Technological Works Published (kind)
合　计 Total	**14 451**	**3 200**	**248**
海洋基础科学研究 Marine Basic Scientific Research	**10 093**	**2 728**	**160**
海洋自然科学 Marine Natural Science	7 202	2 507	85
海洋社会科学 Marine Social Science	583	22	47
海洋农业科学 Marine Agricultural Science	2 289	199	28
海洋生物医药 Marine Biomedicine	19	0	0
海洋工程技术研究 Marine Engineering Technology esearch	**4 088**	**374**	**77**
海洋化学工程技术 Marine Chemical Engineeing Technology	757	82	5
海洋生物工程技术 Marine Bioengineering Technology	0	0	0

6-6 续表 continued

行 业 Industry	发表科技论文（篇） Scientific Theses Published (piece)	#国外发表 Published Abroad	出版科技著作（种） Scientific and Technological Works Published (kind)
海洋交通运输工程技术 Marine Communications and Transport Technology	585	37	6
海洋能源开发技术 Marine Energies Development Technology	980	62	13
海洋环境工程技术 Marine Environmental Engineering Technology	187	21	3
河口水利工程技术 Eustuarine Water Conservancy Engineering	1 216	153	38
其他海洋工程技术 Other Marine Engineering Technology	363	19	12
海洋信息服务业 **Marine Information Service**	**205**	**83**	**11**
其他海洋信息服务 Other Marine Information Services	205	83	11
海洋技术服务业 **Marine Technological Service Industry**	**65**	**15**	**0**
其他海洋专业技术服务 Other Marine Professional and Technological Services	23	11	0
海洋工程管理服务 Marine Engineering Management Service	42	4	0

6-7 分行业科研机构科技专利情况
Marine Scientific and Technological Patents of the Research Institutions by Industry

单位：件 (number)

行 业 Industry	专利申请受理数 Number of Patent Applications Accepted	#发明专利 Patents for Discoveries	专利授权数 Number of Patents Granted	#发明专利 Patents for Discoveries	拥有发明专利总数 Total Number of Patents for Discoveries
合 计 Total	**2 550**	**2 160**	**1 250**	**926**	**6 244**
海洋基础科学研究 Marine Basic Scientific Research	**969**	**785**	**552**	**367**	**1 687**
海洋自然科学 Marine Natural Science	730	639	452	332	1 572
海洋社会科学 Marine Social Science	0	0	0	0	0
海洋农业科学 Marine Agricultural Science	233	141	100	35	106
海洋生物医药 Marine Biomedicine	6	5	0	0	9
海洋工程技术研究 Marine Engineering Technology Research	**1 568**	**1 364**	**692**	**557**	**4 545**
海洋化学工程技术 Marine Chemical Engineeing Technology	1 260	1 206	504	475	4 138
海洋生物工程技术 Marine Bioengineering Technology	0	0	0	0	0

6-7 续表 continued

行　业 Industry	专利申请受理数 Number of Patent Applications Accepted	#发明专利 Patents for Discoveries	专利授权数 Number of Patents Granted	#发明专利 Patents for Discoveries	拥有发明专利总数 Total Number of Patents for Discoveries
海洋交通运输工程技术 Marine Communications and Transport Technology	67	33	25	12	31
海洋能源开发技术 Marine Energies Development Technology	110	68	75	24	193
海洋环境工程技术 Marine Environmental Engineering Technology	10	7	4	3	13
河口水利工程技术 Eustuarine Water Conservancy Engineering Technology	76	25	67	34	143
其他海洋工程技术 Other Marine Engineering Technology	45	25	17	9	27
海洋信息服务业 Marine Information Service	**0**	**0**	**0**	**0**	**0**
其他海洋信息服务 Other Marine Information Services	0	0	0	0	0
海洋技术服务业 Marine Technological Service Industry	**13**	**11**	**6**	**2**	**12**
其他海洋专业技术服务 Other Marine Professional and Technological Services	13	11	5	1	11
海洋工程管理服务 Marine Engineering Management Service	0	0	1	1	1

6-8 分地区海洋科研机构及人员情况
Marine Scientific Research Institutions and Personnel by Regions

地　区 Region	机构数（个） Number of Institutions (unit)	从业人员（人） Employed Population (person)
合 计 Total	**186**	**34 076**
北　京 Beijing	25	12 115
天　津 Tianjin	15	2 491
河　北 Hebei	5	542
辽　宁 Liaoning	17	1 813
上　海 Shanghai	15	3 399
江　苏 Jiangsu	12	2 902
浙　江 Zhejiang	18	1 410
福　建 Fujian	12	1 051
山　东 Shandong	22	3 466
广　东 Guangdong	28	2 690
广　西 Guangxi	9	433
海　南 Hainan	3	192
其　他 Other	5	1 572

6-9 分地区海洋科研机构科技活动人员学历构成

Educational Background Composition of the Personnel Engaged in Scientific and Technological Activities in Marine Scientific Research Institutions by Regions

单位：人 (person)

地 区 Region	科技活动人员 Personnel Engaged in Scientifical Activities	研究生 Postgraduate	#博士 Doctor	大学生 Graduate	大专生 College Graduates	其 他 Others
合 计 Total	**27 888**	**11 453**	**4 599**	**9 866**	**4 200**	**2 369**
北 京 Beijing	10 026	5 099	2 420	2 843	1 464	620
天 津 Tianjin	1 860	578	77	903	205	174
河 北 Hebei	520	120	18	266	84	50
辽 宁 Liaoning	1 583	324	59	759	281	219
上 海 Shanghai	2 906	1 095	365	1 087	478	246
江 苏 Jiangsu	2 023	644	214	634	489	256
浙 江 Zhejiang	1 171	343	86	551	186	91
福 建 Fujian	939	294	59	432	122	91
山 东 Shandong	2 882	1 103	470	1 042	434	303
广 东 Guangdong	2 162	1 002	476	745	245	170
广 西 Guangxi	321	54	7	181	55	31
海 南 Hainan	173	28	3	61	22	62
其 他 Other	1 322	769	345	362	135	56

6-10 分地区海洋科研机构科技活动人员职称构成
Technical Title Composition of Personel Engaged in Marine Scientific Research Institutions by Regions

单位：人 (person)

地 区 Region	科技活动人员 Personnel Engaged in Scientifical Activities	高级职称 Senior	中级职称 Intermediate	其 他 Others
合 计 **Total**	**27 888**	**10 398**	**8 694**	**8 796**
北 京 Beijing	10 026	4 265	3 219	2 542
天 津 Tianjin	1 860	630	575	655
河 北 Hebei	520	181	114	225
辽 宁 Liaoning	1 583	577	428	578
上 海 Shanghai	2 906	948	970	988
江 苏 Jiangsu	2 023	678	475	870
浙 江 Zhejiang	1 171	439	401	331
福 建 Fujian	939	308	296	335
山 东 Shandong	2 882	992	809	1 081
广 东 Guangdong	2 162	795	699	668
广 西 Guangxi	321	58	119	144
海 南 Hainan	173	17	26	130
其 他 Other	1 322	510	563	249

6-11 分地区海洋科研机构经费收入
Routine Fund Receipts of Marine Scientific Research Institutions by Regions

单位：万元 (10 000 yuan)

地 区 Region	经费收入总额 Fund Total	经常费 Routine Fund	科技活动借贷款 Loans in the scientific and technological activities	基本建设中政府投资 Government investment in the capital construction
合 计 Total	**1 601 610**	**1 487 639**	**1 334**	**112 637**
北 京 Beijing	637 946	614 348	0	23 598
天 津 Tianjin	129 653	124 940	0	4 713
河 北 Hebei	6 865	6 815	0	50
辽 宁 Liaoning	65 880	65 790	0	90
上 海 Shanghai	196 031	177 192	0	18 839
江 苏 Jiangsu	63 944	59 747	0	4 196
浙 江 Zhejiang	68 794	68 081	0	713
福 建 Fujian	42 865	42 741	0	124
山 东 Shandong	184 747	148 988	1 300	34 460
广 东 Guangdong	134 119	111 080	0	23 040
广 西 Guangxi	7 517	7 329	0	188
海 南 Hainan	4 040	3 640	0	400
其 他 Other	59 209	56 947	34	2 228

6-12 分地区海洋科研机构科技课题情况

Marine Science and Technology Research Projects of the Research Institutions by Regions

单位：项 (item)

地　区 Region	课题数 Number of research subjectss	基础研究 Basic Research	应用研究 Applied Research	试验发展 Experimental Development	成果应用 Result Application	科技服务 Scientific and Technological Service
合　计 Total	**12 600**	**2 753**	**3 338**	**2 550**	**1 074**	**2 885**
北　京 Beijing	4 402	1 072	1 033	896	187	1 214
天　津 Tianjin	526	0	25	205	80	216
河　北 Hebei	57	2	11	9	16	19
辽　宁 Liaoning	242	0	15	109	91	27
上　海 Shanghai	1 040	122	396	228	77	217
江　苏 Jiangsu	1 434	62	493	381	240	258
浙　江 Zhejiang	536	70	93	60	79	234
福　建 Fujian	620	138	134	110	79	159
山　东 Shandong	1 254	357	450	250	83	114
广　东 Guangdong	1 519	446	447	183	63	380
广　西 Guangxi	100	12	24	24	30	10
海　南 Hainan	50	0	3	0	41	6
其　他 Other	820	472	214	95	8	31

6-13 分地区海洋科研机构科技论著情况
Marine Scientific and Technological Works of the Research Institutions by Regions

地 区 Region	发表科技论文（篇） Scientific Theses Published (piece)	#国外发表 Published Abroad	出版科技著作（种） Scientific and Technological Works Published (kind)
合 计 **Total**	**14 451**	**3 200**	**248**
北 京 Beijing	6 370	1 672	91
天 津 Tianjin	548	28	10
河 北 Hebei	490	0	38
辽 宁 Liaoning	243	29	3
上 海 Shanghai	851	189	10
江 苏 Jiangsu	933	161	15
浙 江 Zhejiang	472	69	11
福 建 Fujian	359	87	1
山 东 Shandong	1 619	421	26
广 东 Guangdong	1 260	342	21
广 西 Guangxi	86	2	1
海 南 Hainan	45	0	0
其 他 Other	1 175	200	21

6-14 分地区海洋科研机构科技专利情况
Marine Scientific and Technological Patents of the Research Institutions by Regions

单位：件 (number)

地 区 Region	专利申请受理数 Number of Patent Applications Accepted	#发明专利 Patents for Discoveries	专利授权数 Number of Patents Granted	#发明专利 Patents for Discoveries	拥有发明专利总数 Total Number of Patents for Discoveries
合 计 Total	**2 550**	**2 160**	**1 250**	**926**	**6 244**
北 京 Beijing	1 160	1 049	557	453	3 439
天 津 Tianjin	70	32	42	21	56
河 北 Hebei	2	0	2	0	6
辽 宁 Liaoning	321	289	118	110	895
上 海 Shanghai	498	414	213	162	873
江 苏 Jiangsu	55	34	28	5	37
浙 江 Zhejiang	31	22	17	9	30
福 建 Fujian	21	13	6	0	3
山 东 Shandong	180	154	128	100	411
广 东 Guangdong	167	129	105	51	376
广 西 Guangxi	1	0	0	0	0
海 南 Hainan	0	0	2	2	2
其 他 Other	44	24	32	13	116

主要统计指标解释

1. 海洋科研机构 指有明确的研究方向和任务，有一定水平的学术带头人和一定数量、质量的研究人员，有开展研究工作的基本条件，长期有组织地从事海洋研究与开发活动的机构。

2. 从业人员 指由本机构年末直接组织安排工作并支付工资的各类人员总数。包括固定职工、国家有编制的合同制职工、招聘人员和返聘的离退休人员。不包括离退休人员、停薪留职人员。

3. 从事科技活动人员 指从业人员中的科技管理人员、课题活动人员和科技服务人员。

4. 高级职称 指研究员、副研究员；教授、副教授；高级工程师；高级农艺师；正、副主任医（药、护、技)师；高级实验师；高级统计师；高级经济师；高级会计师；编审(正、副编审)；译审(正、副译审)、高级(主任)记者；正、副研究馆员等。

5. 中级职称 指助理研究员；讲师；工程师；农艺师；主治医(药、护、技)师；实验师；统计师；经济师；会计师；编辑；翻译；记者；馆员等。

6. 初级职称 指研究实习员；助教；助理工程师、技术员；助理农艺师、农业技术员；医(药、护、技)师、医(药、护、技)士；助理实验师、实验员；助理统计师、统计员；助理经济师；助理会计师、会计员；助理编辑、见习编辑；助理翻译；助理记者；助理馆员、管理员等。

7. 科技经费筹集额 指从各种渠道筹集到的计划用于本单位科技活动的经费，不论来源渠道如何。

8. 政府资金 指由各级政府部门直接拨款或企事业单位利用政府资金委托本机构从事科学技术活动所获得的收入。

9. 生产经营活动收入 指本机构在科研、技术等专业业务活动以外开展非独立核算的经营活动取得的收入，包括产品（商品）销售收入、经营服务收入、工程承包收入、租赁收入和其他经营收入。

10. 其他收入 指开展科技活动与生产经营活动以外的各项活动的收入，包括：用于离退休人员的政府拨款。

11. 非科技活动借贷款 指本机构为开展非科技活动从各种渠道获得的各类借、贷款。不论偿还形式、期限和数额如何，均按当年获得的借、贷款额填报。不包括基本建设贷款。

12. 基础研究 为获得新知识而进行的独创性研究。其目的是揭示观察到的现象和事实的基本原理和规律，而不以任何特定的实际应用为目的。

13. 应用研究 为获得新的科学技术知识而进行的独创性研究。它主要针对某一特定的实际应用目的。应用研究通常是为了确定基础研究成果或知识的可能的用途，或是为达到某一具体的、预定的实际目的确定新的方法(原理性)或途径。

14. 试验发展 利用从研究或实际经验获得的知识，为生产新的材料、产品和装置，建立新的工艺和系统，以及对已生产或建立的上述各项进行实质性的改进而进行的系统性工作。

15. 成果应用 为解决 R&D 活动阶段产生的新产品、新装置、新工艺、新技术、新方法、新系统和服务等能投入生产或在实际应用中所存在的技术问题而进行的系统性活动。它不具有创新成分。此类活动包括为达到生产目的而进行的定型设计和试制以及为扩大新产品的生产规模和探索新方法、新技术、新工艺等的应用领域而进行的适应性试验。

16. 科技服务 与科学研究与实验发展有关，并有助于科学技术知识的产生、传播和应用的活动。包括为扩大科技成果的使用范围而进行的示范性推广工作；为用户提供科技信息和文献服务的系统性工作；为用户提供可行性报告、技术方案、建议及进行技术论证等技术咨询工作；自然、生物现象的日常观测、监测，资源的考查和勘探；有关社会、人文、经济现象的通用资料的收集，如统计、市场调查等以及这些资料的常规分析与整理；为社会和公众提供的测试、标准化、计量、计算、质量控制和专利服务，不包括工商企业为进行正常生产而开展的上述活动。

17. 生产性活动 由于业务特殊的工艺设备条件，或掌握某种技术专长或诀窍，所进行的小量非常规生产。

18. 科技论文 在全国性学报或学术刊物上、省部属大专院校对外正式发行的学报或学术刊物上发表的论文以及向国外发表的论文。

19. 科技著作 经过正式出版部门编印出版的科技专著、大专院校教科书、科普著作。

20. 专利申请受理数 当年本单位向专利管理部门提出申请并被受理的职务专利申请件数。

21. 专利授权数 当年由专利管理部门授予本单位专利权的职务专利件数。

Explanatory Notes on Main Statistical Indicators

1. Marine Scientific Research Institution refers to the institution which has definite research orientations and tasks, high-level academic leading personnel and fair-sized, qualified research personnel, and basic conditions for research work and which is engaged for a long time in the marine research and development activities in an organized way.

2. Employees refer to the total number of personnel of various kinds employed and paid by the institution at the end of the year, including fixed employees, contract workers of staff belonging to the state authorized staff , recruited personnel, reemployed retired personnel, but not including the retired and the personnel on leave with pay suspension.

3. Personnel Engaged in Scientific and Technological Activities refers to the personnel for scientific and Technological management, personnel engaged in the activities of research topics and scientific and technological service personnel.

4. Senior Technical Title refer to research scientist, associate research scientist; professor, associate professor; senior engineer; senior agronomist; professor-rank and associate professor-rank doctor (pharmacists, nurses and technicians); senior laboratory technician; senior statisticians; senior economic engineer; chief accountant; senior editor (professor and associate professor ranks); senior translator (professor and associate professor ranks); senior journalist; research librarian (professor and associate professor ranks), etc.

5. Intermediate Technical Title refer to assistant research scientist; lecturer; engineer; agronomist; lecturer-rank doctor (pharmacist, nurse and technician); laboratory technician; lecturer-rank statistician;

economic engineer; accountant; editor; translator; journalist; librarian, etc.

6. Primary Technical Title refers to trainee researcher; assistant; assistant engineer, technician, assistant agronomist, agricultural technician; assistant-rank doctor (pharmacist, nurse and technician); assistant laboratory technician; assistant statistician; assistant economic engineer, assistant accountant; assistant editor, editor on probation; assistant translator; assistant journalist; assistant research librarian, librarian ,etc.

7. Amount of Scientific and Technological Funds Raised refers to the funds raised through all channels planned to be used as funds for the scientific and technological activities in the institution regardless of their source and channels.

8. Funds from government refers to the direct appropriations by the government departments at all levels or the earnings from conducting scientific and technological activities entrusted to the institution by enterprises as institutions by earning the funds from government.

9. Earnings from Production as Business Activities refer to the incomes obtained from the non-independent accounting business activities carried out by the institution beyond the scientific research, technological and professional activities, including these from sale of products(goods), business and service, contracted projects, leasing and other business.

10. Other Incomes refer to those from the various activities conducted other than scientific and technological activities and production and business activities, including the government appropriations for retired personnel.

11. Loan for Non-scientific as Technological Activities refers to the various types loan obtained by the institutions through all channels for carrying out non-scientific and technological activities, not including the load for capital construction. The loan, irrespective of its form of reimbursement, term and amount is filled in a form and submitted to the authorities as the amount acquired in the current year.

12. Basic Research refers to the original research to acquire new knowledge. It is aimed at revealing the basic principles and laws of the phenomena and facts observed, but not at any specific practical applications.

13. Applied Research refers to the original research to acquire new scientific and technological knowledge. It mainly serves the purpose of a particular practical application. The purpose of applied research is usually to define the potential uses of the research finds or knowledge obtained from basic research or to identify new methods (principles) or ways to reach a specific and predetermined goal.

14. Experimental Development refers to the systematic work carried out to establish new technologies and systems for producing new materials, products and equipment by using the knowledge obtained from research or practical experience, or to make substantial improvement of the above-mentioned which have been produced or established.

15. Result Application refers to the systematic activities carried out to solve the technical problems that might crop up in the production or practical application of the new products, devices, technologies, techniques, methods, systems and service occurring in the course of R&D activities.

They do not bring forth new ideas. Such activities include the finalizing design and trial-production for the purpose of production as well as the adaptive tests to expand the production scale of new products and the application areas of new methods, techniques and technologies.

16. Scientific and Technological Service refers to the activities that are associated with the scientific research and experimental development, and contribute to the generation, dissemination and application of scientific and technological knowledge, which include the demonstrative work of popularization to enlarge the use scope of scientific and technological achievements; the systematic work of providing the users with scientific and technological information and literature service; the technical consultation work of providing users with feasibility reports, technical schemes and recommendations and carrying out technical demonstration; routine observation and monitoring of natural and biological phenomena, and the survey and exploration of resources; collection of universal data on the appropriate social, cultural and economic phenomena, such as statistics and market survey, as well as the routine analysis and sorting-out of these data; the provision for the society and the public of such service as testing, standardization, computation, quality control and patent, but not including the type of the above-mentioned activities carried out by industrial and commercial enterprises for the purpose of normal production.

17. Productive Activity refers to the small-scale and non-conventional productions due to the presence of special technologies and equipment or mastery of a particular technical expertise or secret of success.

18. Scientific Treatises refer to the theses published in the national journals or academic publications, those officially issued journals or academic publications by universities and colleges under provinces or ministries as well as theses published abroad.

19. Scientific and Technological Works refer to the scientific and technological monographs, textbooks for universities and colleges and popular science books published by the official publishing houses.

20. Number of Patents Applied and Accepted refers to the number of professional patent applications of the unit to the patent administrative department and accepted by it in the year.

21. Number of Patents Granted refers to the number of the professional patents granted to the unit by the patent administrative department in the year.

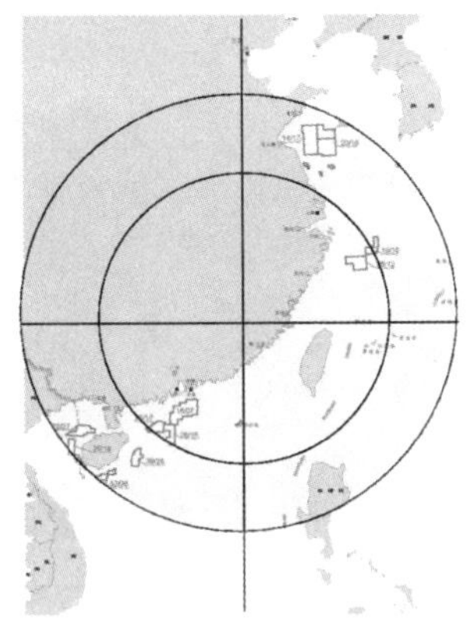

7

海 洋 教 育

Marine Education

7-1 全国各海洋专业博士研究生情况
DoctoralStudentsfromMarineSpecialities

专 业 Speciality	专业点数（个） Number of Speciality Agencies (unit)	学生数（人） Students (person)			
		毕业生 Graduated	招 生 Entrants	在校生 Enrollment	毕业班学生 Graduation
合 计 **Total**	**121**	**627**	**795**	**3 315**	**1 611**
物理海洋学 Physical Oceanography	5	43	64	188	97
海洋化学 Marine Chemistry	4	26	36	129	64
海洋生物学 Marine Biology	9	108	75	274	124
海洋地质 Marine Geology	8	45	46	209	120
海洋科学类新专业 New Speciality of Marine Science	3	38	73	277	139
港口海岸及近海工程 Coastal Harbour and Offshore Engineering	11	37	58	269	111
船舶与海洋结构物设计制造 Ship and Marine Structures Design and Manu facture	10	44	63	374	171
轮机工程 Turbine Engineering	9	12	44	228	77

7-1 续表 continued

专 业 Speciality	专业点数（个） Number of Speciality Agencies (unit)	学生数（人） Students (person)			
		毕业生 Graduated	招 生 Entrants	在校生 Enrollment	毕业班学生 Graduation
水声工程 Hydroacoustic Engineering	6	18	31	157	93
船舶与海洋工程新专业 New Speciality of Shipand Marine Engineering	2	0	3	15	9
捕捞学 Science of Fishing	2	0	4	11	7
航空、航天与航海医学 Aeronautical, Aerospace and Nautical Medicine	1	0	15	15	0
水产品加工及贮藏工程 Aquatic Product Sprocessing and Storing Engineering	6	6	7	40	23
水产新专业 New Specialities of Aquaculture	3	13	10	60	37
水产养殖 Aquaculture	7	32	53	171	76
水生生物学 Hydrobiology	16	69	90	316	144
水文学及水资源 Hydrology and Water Resource	16	133	108	531	293
渔业资源 Fishery Resource	3	3	15	51	26

7-2 全国各海洋专业硕士研究生情况
PostgraduateStudentsfromMarineSpecialities

专业 Speciality	专业点数（个） Number of Speciality Agencies (unit)	学生数（人）Students (person)			
		毕业生 Graduated	招生 Entrants	在校生 Enrollment	毕业班学生 Graduation
合　计 Total	**288**	**2 644**	**3 633**	**10 052**	**3 126**
物理海洋学 Physical Oceanography	10	88	121	309	78
海洋化学 Marine Chemistry	14	80	113	333	104
海洋生物学 Marine Biology	25	169	420	1 063	302
海洋地质 Marine Geology	15	100	132	352	108
海洋科学类新专业 New Speciality of Marine Science	3	24	69	168	39
港口海岸及近海工程 Coastal Harbour and Offshore Engineering	22	248	296	868	263
船舶与海洋结构物设计制造 Shipand Marine Structures Design	18	341	424	1 189	342
轮机工程 Turbine Engineering	14	230	306	787	293

7-2 续表 continued

专 业 Speciality	专业点数（个） Number of Speciality Agencies (unit)	学生数（人）Students (person)			
		毕业生 Graduated	招 生 Entrants	在校生 Enrollment	毕业班学生 Graduation
水声工程 Hydroacoustic Engineering	11	150	147	449	155
船舶与海洋工程新专业 New Specialities in Shipping and Marine Engineering	4	26	13	63	18
捕捞学 Science Offishing	4	18	35	93	30
航空、航天与航海医学 Aeronautical, Aerospace and Nautical Medicine	3	21	14	45	18
水产品加工及贮藏工程 Aquatic Products Processing and Storing Engineering	19	99	90	285	112
水产新专业 New Specialities of Aquaculture	5	16	22	66	22
水产养殖 Aquaculture	28	379	449	1 294	447
水生生物学 Hydrobiology	35	180	400	1 005	259
水文学及水资源 Hydrology and Water Resource	50	394	497	1 419	436
渔业资源 Fishery Resource	8	81	85	264	100

7-3 全国普通高等教育各海洋专业本、专科学生情况
Undergraduates and Students from Colleges for Professional Training from Marine Specialities in the Ordinary National Higher Education

专业 Speciality	专业点数（个） Number of Speciality Agencies (unit)	学生数（人） Students (person)			
		毕业生 Graduated	招生 Entrants	在校生 Enrollment	毕业班学生 Graduation
合计 Total	**590**	**37 245**	**49 699**	**160 717**	**46 725**
海洋科学 Marine Science	20	464	781	2 861	607
海洋技术 Marine Technology	15	404	615	2 033	404
海洋管理 Marine Management	4	43	92	280	43
海洋生物资源与环境 Marine Living Resources and Environment	3	59	105	450	54
海洋科学类新专业 New Speciality of Marine Science	1	0	84	84	0
港口海岸及治河工程 Coastal Harbour and River Control Engineering	5	45	213	654	161
港口航道与海岸工程 Harbour Channel and Coastal Engineering	22	1 061	1 316	4 925	1 147
水资源与海洋工程 Water Resource and Ocean Engineering	2	28	0	133	51
航海技术 Nautical Technology	14	1 547	2 163	8 349	1 899
轮机工程 Turbine Engineering	20	1 794	2 736	10 173	2 196

7-3 续表 1 continued

专 业 Speciality	专业点数（个） Number of Speciality Agencies (unit)	学生数（人）Students (person)			
		毕业生 Graduated	招 生 Entrants	在校生 Enrollment	毕业班学生 Graduation
海事管理 Maritime Affairs Manegernent	2	118	119	508	137
船舶与海洋工程 Ship and Marine Engineering	24	1 819	2 746	10 681	2 240
海洋工程类新专业 New Speciality of Marine Engineering	2	0	307	617	0
水产养殖学 Aquaculture	46	2 055	2 463	9 345	2 148
海洋渔业科学与技术 Science and Technology of Marine Fishery	9	293	409	1 326	284
水族科学与技术 Science and Technology of Aquatic Animals	3	205	227	749	154
水产类新专业 New Speciality in Aquatic Products	1	38	0	115	74
报关与国际货运 Customs Clearing and International Freight Transport	156	11 847	14 329	46 642	15 972
船舶工程技术 Ship Engineering	30	2 808	5 214	15 095	4 170
船舶检验 Ship Inspection	2	108	163	413	132
船舶舾装 Ship Equipment and Installations	2	1	75	75	0
船机制造与维修 Ship Engines Manufacturing and Maintenance	3	0	211	348	0

7-3 续表 2 continued

专 业 Speciality	专业点数（个） Number of Speciality Agencies (unit)	学生数（人）Students (person)			
		毕业生 Graduated	招 生 Entrants	在校生 Enrollment	毕业班学生 Graduation
船艇动力管理 Ship and Boat Power Equipment Management	1	72	156	273	77
港口工程技术 Harbour Engineering	7	207	458	1 152	330
港口机械应用技术 Applied Harbour Machinery Technology	2	227	163	841	382
港口物流设备与自动控制 Harbour Logistics Equipment and Automatic Control	19	1 254	1 216	4 103	1 384
港口业务管理 Harbour Business Management	18	1 216	978	3 099	1 302
港口与航运管理 Harbour and Shipping Management	5	210	347	993	333
港口运输类新专业 New Specialities in Harbour Transportation	3	87	214	374	103
国际航运业务管理 International Shipping Business Management	27	1 712	1 908	5 852	2 146
海关管理 Customs Management	7	493	210	989	473
海关国际法律条约与公约 Customs International Legal treaties and Conventions	1	38	0	50	50
航道工程技术 Channel Engineering	1	0	43	43	0
集装箱运输管理 Container Transportation Management	19	954	1 105	3 386	1 072

7-3 续表 3 continued

专 业 Speciality	专业点数（个） Number of Speciality Agencies (unit)	学生数（人）Students (person)			
		毕业生 Graduated	招 生 Entrants	在校生 Enrollment	毕业班学生 Graduation
轮机工程技术 Engine Engineering	36	4 224	5 843	16 947	5 241
水产养殖技术 Aquaculture Technology	27	881	1 455	3 555	1 054
水产养殖类新专业 New Specialities in Aquaculture	2	66	76	250	74
水环境监测与保护 Water Environmental Monitoring and Protection	6	166	231	635	168
水上运输类新专业 New Specialities in Waterborne Communications	6	304	547	1 101	214
水生动植物保护 Aquatic Animals and Plants Protection	3	86	78	235	73
水文与水资源类 Hydrology and Water Resources	1	0	0	28	28
水文与水资源类新专业 New Specialities in Hydrology and Water Resources	1	65	100	335	106
水信息技术 Hydrological Information Technology	1	35	0	120	83
水运管理 Water Transport Management	2	122	55	249	98
水政水资源管理 Water Resources Management by Water Administration	1	0	0	20	20
特种水产养殖 Special Aquaculture	1	1	0	0	0
渔业综合技术 Integrated Fishery Technologies	7	88	148	231	41

7-4 全国成人高等教育各海洋专业学生情况
Students from Marine Specialities in the National Adult Higher Education

专 业 Speciality	专业点数（个） Number of Speciality Agencies (unit)	学生数（人）Students (person)			
		毕业生 Graduated	招 生 Entrants	在校生 Enrollment	毕业班学生 Graduation
合 计 Total	**133**	**9 924**	**9 379**	**28 006**	**11 745**
海洋科学 Marine Science	1		6	54	34
水文与水资源工程 Hydrological and Water Resources Engineering	24	761	488	1 862	865
港口航道与海岸工程 Harbour Channel and Coastal Engineering	3	27	44	197	111
港口海岸及治河工程 Harbour Coastal and river-harnessing Engineering	1		5	93	66
航海技术 Navigation Technology	25	4 355	3 478	10 273	4 438
轮机工程 Turbine Engineering	29	3 478	2 744	8 532	4 049
海事管理 Maritime Affairs Management	4	160	58	267	167
船舶与海洋工程 Ship and Marine Engineering	17	563	2 124	4 783	1 083
海洋工程类新专业 New Speciality of Marine Engineering	2	24	87	374	86
水产养殖学 Aquaculture	16	473	314	1 141	565
海洋渔业科学与技术 Science and Technology of Marine Fishery	2	30	13	166	153
水产类新专业 New Speciality in Aquatic Products	5	32	18	102	53
航运管理 Shipping Management	4	21		162	75

7-5 全国中等职业教育各海洋专业学生情况
Students from Marine Specialities in the National Secondary Vocational Education

专业 Speciality	专业点数（个） Number of Speciality Agencies (unit)	学生数（人）Students (person)			
		毕业生 Graduated	招生 Entrants	在校生 Enrollment	毕业班学生 Graduation
合计 Total	**382**	**16 774**	**33 487**	**73 446**	**24 372**
水产养殖 Aquaculture	69	1 836	5 403	9 520	2 158
航海捕捞 Marine Fishing	7	370	210	668	381
水文与水资源 Hydrology and Water Resources	1		1	1	1
海洋观测 Marine Observation	2		60	60	
港口与航道工程技术 Harbour and Channel Engineering	6	123	177	424	78
船体建造与修理 Ship Hull Building and Repair	48	2 106	2 419	7 363	2 129
船舶机械装置 Ship Mechanical Equipment	18	697	910	2 197	671
船舶电气技术 Ship Electrical Technology	18	442	796	2 526	903
船舶驾驶 Ship Piloting	88	5 014	12 313	25 344	9 491
轮机管理 Engines Management	60	3 986	9 162	19 740	7 086
船舶水手与机工 Ship Sailors and Mechanics	44	1 455	1 374	2 915	745
外轮理货 Foreign Ships Freight Forwarding	8	274	178	1 084	282
船舶检验 Ship Inspection	3	40	76	305	74
工程潜水 Engineering Diving	3	70	18	54	16
船舶电子设备 Ship Electronic Equipment	3	259	4	278	84
船舶通信与导航 Ship Communications and Navigation	4	102	386	967	273

7-6 分地区海洋专业博士研究生情况
Doctoral Students in Marine Specialities by Regions

地 区 Region	专业点数（个） Number of Speciality Agencies (unit)	学生数（人）Students (person)			
		毕业生 Graduated	招 生 Entrants	在校生 Enrollment	毕业班学生 Graduation
合 计 Total	**121**	**627**	**795**	**3 315**	**1 611**
北 京 Beijing	5	16	23	89	38
天 津 Tianjin	3	10	0	24	16
辽 宁 Liaoning	5	42	52	319	41
上 海 Shanghai	17	38	86	288	131
江 苏 Jiangsu	13	73	69	394	253
浙 江 Zhejiang	3	0	7	21	3
福 建 Fujian	6	15	27	109	58
山 东 Shandong	19	224	270	1 029	554
广 东 Guangdong	11	75	69	217	86
其 他 Others	39	134	192	825	431

7-7 分地区海洋专业硕士研究生情况
Postgraduate Students in Marine Specialities by Regions

地 区 Region	专业点数（个） Number of Speciality Agencies (unit)	学生数（人）Students (person)			
		毕业生 Graduated	招 生 Entrants	在校生 Enrollment	毕业班学生 Graduation
合 计 Total	**288**	**2 644**	**3 633**	**10 052**	**3 126**
北 京 Beijing	18	91	119	313	95
天 津 Tianjin	7	71	34	162	37
河 北 Hebei	5	17	19	61	22
辽 宁 Liaoning	20	342	357	1 028	393
上 海 Shanghai	27	317	460	1 182	355
江 苏 Jiangsu	23	294	379	1 095	360
浙 江 Zhejiang	20	109	181	476	138
福 建 Fujian	15	92	153	408	120
山 东 Shandong	35	271	539	1 416	384
广 东 Guangdong	24	195	279	769	234
广 西 Guangxi	2	8	24	67	20
海 南 Hainan	2	23	27	71	22
其 他 Others	90	814	1 062	3 004	946

7-8 分地区普通高等教育海洋专业本、专科学生情况
Undergraduates and Students from Colleges for Professional Training in the Marine Specialities of Ordinary Higher Education by Regions

地　区 Region	专业点数（个） Number of Speciality Agencies (unit)	学生数（人）Students (person)			
		毕业生 Graduated	招 生 Entrants	在校生 Enrollment	毕业班学生 Graduation
合　计 Total	**590**	**37 245**	**49 699**	**160 717**	**46 725**
北　京 Beijing	3	56	160	534	58
天　津 Tianjin	24	2 310	2 232	7 557	2 154
河　北 Hebei	30	1 190	1 736	5 121	1 566
辽　宁 Liaoning	35	2 693	3 986	13 509	3 691
上　海 Shanghai	51	4 008	4 140	14 072	4 370
江　苏 Jiangsu	68	4 799	5 952	21 619	6 634
浙　江 Zhejiang	43	1 864	2 472	8 790	2 934
福　建 Fujian	39	2 557	3 124	10 201	3 158
山　东 Shandong	83	5 274	8 015	24 575	6 857
广　东 Guangdong	36	1 986	2 807	7 992	2 072
广　西 Guangxi	18	514	1 529	3 512	904
海　南 Hainan	9	860	734	3 079	1 277
其　他 Others	151	9 134	12 812	40 156	11 050

7-9 分地区成人高等教育海洋专业学生情况
Students from Marine Specialities in the Adult Higher Education by Regions

地区 Region	专业点数（个）Number of Speciality Agencies (unit)	学生数（人）Students (person)			
		毕业生 Graduated	招生 Entrants	在校生 Enrollment	毕业班学生 Graduation
合计 Total	**133**	**9 924**	**9 379**	**28 006**	**11 745**
北京 Beijing	1	94	11	41	30
天津 Tianjin	8	570	537	1 718	956
河北 Hebei	6	175	258	623	212
辽宁 Liaoning	14	3 694	2 520	6 172	3 181
上海 Shanghai	10	492	571	1 872	814
江苏 Jiangsu	10	1 331	1 014	3 451	1 290
浙江 Zhejiang	9	2 213	1 342	5 478	2 220
福建 Fujian	7	182	438	1 467	486
山东 Shandong	17	325	555	1 754	714
广东 Guangdong	9	39	805	1 496	379
广西 Guangxi	2	0	1	5	4
其他 Others	40	809	1 327	3 929	1 459

7-10 分地区中等职业教育海洋专业学生情况
Students from Marine Specialities in the Secondary Vocational Education by Regions

地 区 Region	专业点数（个） Number of Speciality Agencies (unit)	学生数（人） Students (person)			
		毕业生 Graduated	招 生 Entrants	在校生 Enrollment	毕业班学生 Graduation
合 计 Total	**382**	**16 774**	**33 487**	**73 446**	**24 372**
天 津 Tianjin	9	900	874	1 669	776
河 北 Hebei	18	605	932	2 032	365
辽 宁 Liaoning	33	1 427	1 628	3 672	1 445
上 海 Shanghai	12	552	384	1 573	380
江 苏 Jiangsu	53	978	2 505	6 207	858
浙 江 Zhejiang	13	965	724	2 501	836
福 建 Fujian	54	2 674	5 660	12 727	4 535
山 东 Shandong	39	2 248	9 874	19 129	7 736
广 东 Guangdong	14	629	1 162	2 814	1 044
广 西 Guangxi	9	291	674	1 369	259
海 南 Hainan	2	22	152	234	23
其 他 Others	126	5 483	8 918	19 519	6 115

7-11 分地区开设海洋专业高等学校教职工数
Number of Teaching and Administrative Staff in the Universities and Colleges Offering Marine Specialities by Regions

地 区 Region	机构数（个） Number of Institutions (unit)	教职工数（人） Number of Teaching and Administrative Staff (person)	专任教师数（人） Number of Full-Time Teacher (person)
合 计 Total	**314**	**387 857**	**229 988**
北 京 Beijing	3	5 224	3 079
天 津 Tianjin	12	14 735	8 986
河 北 Hebei	19	17 563	10 410
辽 宁 Liaoning	16	14 502	8 499
上 海 Shanghai	18	22 503	11 272
江 苏 Jiangsu	36	52 677	33 013
浙 江 Zhejiang	18	21 908	12 353
福 建 Fujian	15	10 732	7 191
山 东 Shandong	42	49 922	31 061
广 东 Guangdong	14	22 014	13 121
广 西 Guangxi	13	10 057	6 490
海 南 Hainan	5	5 267	3 207
其 他 Others	103	140 753	81 306

主要统计指标解释

海洋专业 指高等教育和中等职业教育所设的与海洋有关的专业。

Explanatory Notes on Main Statistical Indicators

Marine Speciality refers to the ocean-related speciality in high learning and the professional secondary vocational education.

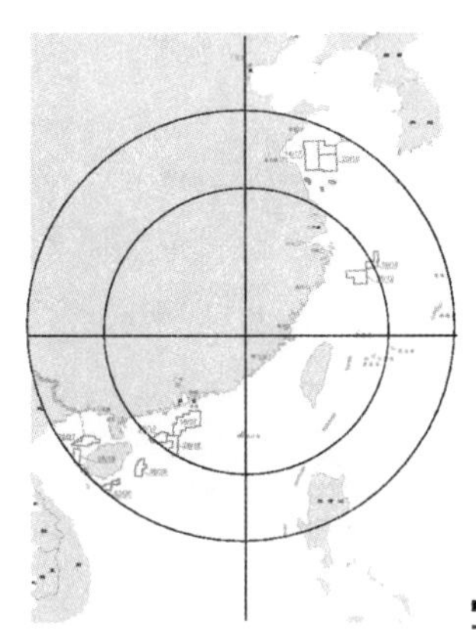

8

海洋环境保护

Marine Environmental Protection

8-1 海区海水质量监测结果
Seawater Quality Monitoring Results

单位：毫克/升，叶绿素a为微克/升 (mg/L Chl a: μg/L)

项　目 Item	渤　海 Bohai Sea				黄　海 Huanghai Sea			
	最大值 Maximum value	最小值 Minimum value	平均值 Mean value	中位值 Median value	最大值 Maximum value	最小值 Minimum value	平均值 Mean value	中位值 Median value
盐　度 Salinity	33. 8340	22. 8894	30. 5068	30. 9490	33. 0810	21. 3200	29. 7970	30. 5200
pH	8. 4100	7. 6300	8. 0962	8. 0900	9. 1300	7. 5100	8. 0910	8. 0400
化学需氧量 COD	4. 2000	0. 3760	1. 3568	1. 1800	3. 8400	0. 1700	1. 0283	0. 9900
溶解氧 DO	9. 1200	3. 1000	6. 6307	6. 8400	10. 6800	1. 3500	6. 9547	6. 9800
石油类 Oils	0. 0860	0. 0032	0. 0335	0. 0295	0. 2130	0. 0031	0. 0380	0. 0255
活性磷酸盐 Active Phosphate	0. 0626	0. 0002	0. 0134	0. 0086	0. 0650	0. 0005	0. 0101	0. 0057
亚硝酸盐氮 Nitrite-nitrogen	0. 3180	0. 0005	0. 0240	0. 0154	0. 2820	0. 0001	0. 0111	0. 0049
硝酸盐氮 Nitrate-nitrogan	1. 4000	0. 0009	0. 1494	0. 0843	1. 9800	0. 0001	0. 1016	0. 0625
氨　氮 Nitrogen Amide	0. 6570	0. 0001	0. 0516	0. 0421	0. 3090	0. 0011	0. 0498	0. 0397
无机氮 Inorganic Nitrogen	1. 5500	0. 0168	0. 2244	0. 1541	2. 0800	0. 0078	0. 1625	0. 1100
叶绿素a Chl a	19. 6500	0. 2370	2. 5662	1. 6641	31. 2000	0. 0320	2. 3031	1. 2700

8-1 续表 continued

项　目 Item	东　海 East China Sea				南　海 South China Sea			
	最大值 Maximum value	最小值 Minimum value	平均值 Mean value	中位值 Median value	最大值 Maximum value	最小值 Minimum value	平均值 Mean value	中位值 Median value
盐　度 Salinity	34. 7200	0. 1480	28. 7609	30. 4500	35. 5000	0. 1300	27. 6643	30. 3800
pH	8. 4900	7. 0300	8. 0301	8. 0400	8. 5700	6. 9900	8. 0990	8. 1300
化学需氧量 COD	2. 6000	0. 0600	0. 7908	0. 7100	4. 7500	0. 0600	0. 8340	0. 5800
溶解氧 DO	10. 0000	2. 3700	6. 1069	6. 2200	12. 0100	1. 1300	6. 1107	6. 3100
石油类 Oils	0. 2060	0. 0010	0. 0215	0. 0170	0. 9040	0. 0010	0. 0383	0. 0210
活性磷酸盐 Active Phosphate	0. 1710	0. 0007	0. 0264	0. 0229	0. 3299	0. 0003	0. 0324	0. 0250
亚硝酸盐氮 Nitrite-nitrogen	0. 0850	0. 0007	0. 0154	0. 0100	0. 3178	0. 0006	0. 0247	0. 0098
硝酸盐氮 Nitrate-nitrogan	2. 0300	0. 0010	0. 3431	0. 1860	2. 7000	0. 0010	0. 1618	0. 0560
氨　氮 Nitrogen Amide	0. 2890	0. 0020	0. 0340	0. 0267	2. 8335	0. 0010	0. 1136	0. 0200
无机氮 Inorganic Nitrogen	2. 0918	0. 0090	0. 3925	0. 2500	3. 1920	0. 0010	0. 2916	0. 0941
叶绿素a Chl a	35. 7000	0. 0100	3. 1236	1. 3700	48. 2000	0. 0100	3. 1351	1. 3300

8-2 海区海水水质评价结果
Seawater Quality Assessment Results

项　目 Item	**全国总计 National Total**	渤　海 Bohai Sea	黄　海 Huanghai Sea	东　海 East China Sea	南　海 South China Sea
测站个数（个） Number of Survey Stations (number)	**845**	159	190	250	246
较清洁海域面积（万平方公里） Cleaner Sea Area ($10\ 000km^2$)	**7.10**	0.90	1.13	3.08	1.99
轻度污染海域面积（万平方公里） Lightly-Polluted Sea Area ($10\ 000km^2$)	**2.55**	0.57	0.79	0.90	0.29
中度污染海域面积（万平方公里） Moderately-Polluted Sea Area ($10\ 000km^2$)	**2.09**	0.42	0.52	0.87	0.28
严重污染海域面积（万平方公里） Heavily-Polluted Sea Area ($10\ 000km^2$)	**2.97**	0.27	0.22	1.96	0.52
首要超标污染物 Prime Pollutants Exceeding the Set Standard	**无机氮**	无机氮	无机氮	无机氮	无机氮

8-3 海区废弃物海洋倾倒情况
Ocean Dumping of Wastes by Sea Area

海　区 Sea Area	疏浚物 （立方米） Dredged Materials (10 000m^3)	惰性无机地质废料 （立方米） Inert, Inorganic Geologic Wastes (m^3)	人体骨灰 （盒） Human Ashes (case)
合　计 **Total**	**95 671 040**	**7 219 159**	
渤黄海 Bohai and Huanghai Sea	49 598 300		
东　海 East China Sea	46 070 000		
南　海 South China Sea	2 740	7 219 159	

8-4 海区海洋石油勘探开发污染物排放入海情况
Discharge of Pollutants into the Sea from Offshore Oil Exploration and Exploitation

海 区 Sea Area	生产污水 （万立方米） Sewage from Production (10 000m^3)	泥浆 （立方米） Sludge (m^3)	钻屑 （立方米） Debris from Drilling (m^3)	机舱污水 （立方米） Sewage from Engineroom (m^3)	食品废弃物 （立方米） Food Wastes (m^3)	生活污水 （立方米） Domestic Sewage (m^3)
合 计 Total		**60 341.72**	**50 960.39**	**7 321.50**	**682.50**	**287 336.40**
渤 海 Bohai Sea	659.12	14 528.72	38 577.39	157.50	54.50	110 516.40
黄 海 Huanghai Sea						
东 海 East China Sea	113.39	831.00	132.00		21.00	23 735.00
南 海 South China Sea	14 709.90	44 982.00	12 251.00	7 164.00	607.00	153 085.00

8-5 沿海地区工业废水排放及处理情况
Discharge and Treatment of Industrial Waste Water by Coastal Regions

单位：万吨 (10 000 tons)

地 区 Region		工业废水排放总量 Total Volume of Industrial Waste Water Discharged		工业废水中 Industrial Waste Water	
			直接排入海的 Discharged Directly to Sea	符合排放标准的 Up-to-the-Standard	达标率（%） Up-to-the-Standard Rate (%)
总 计 Total		**1 388 342.6**	**134 695.1**	**1 333 712.7**	**96.1**
环渤海经济区 Round-the-Bohai Sea Economic Zone	**合 计 Total**	**387 330.3**	**33 675.2**	**372 228.8**	**96.1**
	辽 宁 Liaoning	75 158.6	24 434.5	64 593.0	85.9
	河 北 Hebei	110 058.0	1 162.1	108 165.8	98.3
	天 津 Tianjin	19 441.1	583.6	19 440.1	100.0
	山 东 Shandong	182 672.6	7 495.0	180 029.9	98.6
长江三角洲经济区 Yangzi River Delta Economic Zone	**合 计 Total**	**500 793.7**	**15 799.6**	**485 823.5**	**97.0**
	江 苏 Jiangsu	256 160.0	2 535.9	251 289.6	98.1
	上 海 Shanghai	41 192.0	1 622.7	40 686.6	98.8
	浙 江 Zhejiang	203 441.7	11 641.0	193 847.3	95.3
海峡西岸经济区 Economic Zone on West Side of the Straits	**合 计 Total**	**142 747.0**	**73 810.9**	**141 032.5**	**98.8**
	福 建 Fujian	142 747.0	73 810.9	141 032.5	98.8
珠江三角洲经济区 Zhujiang River Delta Economic Zone	**合 计 Total**	**188 843.9**	**7 549.0**	**174 380.0**	**92.3**
	广 东 Guangdong	188 843.9	7 549.0	174 380.0	92.3
环北部湾经济区 Round-the-Beibu Gulf Economic Zone	**合 计 Total**	**168 627.7**	**3 860.4**	**160 247.9**	**95.0**
	广 西 Guangxi	161 596.4	1 597.0	153 458.4	95.0
	海 南 Hainan	7 031.3	2 263.4	6 789.5	96.6

8-6 沿海城市工业废水排放及处理情况

Discharge and Treatment of Industrial Waste Water by Coastal Cities

单位：万吨 (10 000 tons)

沿海城市 Coastal City	工业废水排放总量 Total Volume of Industrial Waste Water Discharged		工业废水中 Industrial Waste Water	
		直接排入海的 Discharged Directly to Sea	符合排放标准的 Up-to-the-Standard	达标率（%） Up-to-the-Standard Rate (%)
天　津 Tianjin	19 441	584	19 440	100.0
唐　山 Tangshan	20 216	571	19 805	98.0
秦皇岛 Qinhuangdao	5 130	501	5 114	99.7
沧　州 Cangzhou	8 098	89	7 857	97.0
大　连 Dalian	27 820	23 123	26 503	95.3
丹　东 Dandong	5 196	60	2 867	55.2
锦　州 Jinzhou	3 580	78	3 215	89.8
营　口 Yingkou	3 350	507	3 343	99.8
盘　锦 Panjin	1 751		1 494	85.3
葫芦岛 Huludao	2 155	664	1 616	75.0
上　海 Shanghai	41 192	2 536	40 687	98.8
南　通 Nantong	15 943	0	15 928	99.9
连云港 Lianyungang	3 244	343	3 184	98.2
盐　城 Yancheng	10 081	1 008	9 425	93.5

8-6 续表1 continued

沿海城市 Coastal City	工业废水排放总量 Total Volume of Industrial Waste Water Discharged	直接排入海的 Discharged Directly to Sea	工业废水中 Industrial Waste Water	
			符合排放标准的 Up-to-the-Standard	达标率（%） Up-to-the-Standard Rate (%)
杭　州 Hangzhou	79 959	1 451	76 925	96.2
宁　波 Ningbo	17 736	7 757	15 423	87.0
温　州 Wenzhou	7 900	411	6 784	85.9
嘉　兴 Jiaxing	17 488	477	17 218	98.5
绍　兴 Shaoxing	29 862	6	29 351	98.3
舟　山 Zhoushan	1 856	1 198	1 787	96.2
台　州 Taizhou	5 125	342	4 628	90.3
福　州 Fuzhou	4 288	378	3 860	90.0
厦　门 Xiamen	3 776	155	3 755	99.5
莆　田 Putian	1 571	486	1 498	95.4
泉　州 Quanzhou	26 283	1 773	26 240	99.8
漳　州 Zhangzhou	77 058	70 909	76 746	99.6
宁　德 Ningde	2 065	110	2 006	97.1
青　岛 Qingdao	10 402	1 799	10 390	99.9
东　营 Dongying	10 079	0	10 079	100.0
烟　台 Yantai	7 600	2 172	7 600	100.0
潍　坊 Weifang	17 343	0	17 167	99.0
威　海 Weihai	2 252	576	2 252	100.0
日　照 Rizhao	8 826	2 949	8 826	100.0
滨　州 Binzhou	14 097	0	13 473	95.6

8-6 续表2 continued

沿海城市 Coastal City	工业废水排放总量 Total Volume of Industrial Waste Water Discharged	直接排入海的 Discharged Directly to Sea	工业废水中 Industrial Waste Water	
			符合排放标准的 Up-to-the-Standard	达标率（%） Up-to-the-Standard Rate (%)
广 州 Guangzhou	26 023	2 316	25 116	96.5
深 圳 Shenzhen	8 073	606	7 776	96.3
珠 海 Zhuhai	5 891	98	5 774	98.0
汕 头 Shantou	4 972	95	4 491	90.3
江 门 Jiangmen	11 429	345	10 789	94.4
湛 江 Zhanjiang	5 272	197	4 471	84.8
茂 名 Maoming	7 476	914	6 399	85.6
惠 州 Huizhou	5 782	668	5 601	96.9
汕 尾 Shanwei	4 217	3	1 981	47.0
阳 江 Yangjiang	2 426	19	1 900	78.3
东 莞 Dongguan	29 962	2 222	29 310	97.8
中 山 Zhongshan	12 629	41	12 396	98.2
潮 州 Chaozhou	3 922		3 274	83.5
揭 阳 Jieyang	4 407	20	4 198	95.3
北 海 Beihai	1 211	77	1 115	92.1
防城港 Fangchenggang	3 862	583	3 529	91.4
钦 州 Qinzhou	4 106	937	3 949	96.2
海 口 Haikou	475		475	100.0
三 亚 Sanya	25	0	25	100.0

8-7 沿海地带工业废水排放及处理情况
Discharge and Treatment of Industrial Waste Water by Coastal Counties

单位：万吨 (10 000 tons)

地 区 Region	工业废水排放总量 Total Volume of Industrial Waste Water Discharged	直接排入海的 Discharged Directly to Sea	工业废水中 Industrial Waste Water 符合排放标准的 Up-to-the-Standard	达标率（%） Up-to-the-Standard Rate (%)
天 津 Tianjin	**7 444**	**581**	**7 444**	**100.0**
河 北 Hebei	**14 191**	**1 143**	**13 993**	**99.0**
唐 山 Tangshan	6 255	571	6 076	97.0
秦皇岛 Qinhuangdao	4 748	501	4 735	100.0
沧 州 Cangzhou	3 188	71	3 182	100.0
辽 宁 Liaoning	**28 479**	**19 399**	**25 577**	**90.0**
大 连 Dalian	21 295	18 168	20 152	95.0
丹 东 Dandong	1 130	60	112	10.0
锦 州 Jinzhou	1 683	0	1 406	84.0
营 口 Yingkou	2 249	507	2 248	100.0
盘 锦 Panjin	77	0	65	84.0
葫芦岛 Huludao	2 045	664	1 594	78.0
上 海 Shanghai	**29 341**	**2 536**	**28 919**	**99.0**
江 苏 Jiangsu	**15 395**	**393**	**15 031**	**98.0**
南 通 Nantong	6 455	0	6 453	100.0
连云港 Lianyungang	2 783	343	2 731	98.0
盐 城 Yancheng	6 157	50	5 847	95.0

8-7 续表1 continued

地 区 Region	工业废水排放总量 Total Volume of Industrial Waste Water Discharged	直接排入海的 Discharged Directly to Sea	工业废水中 Industrial Waste Water 符合排放标准的 Up-to-the-Standard	达标率（%） Up-to-the-Standard Rate (%)
浙 江 Zhejiang	**78 716**	**9 265**	**74 485**	**95.0**
杭 州 Hangzhou	24 310	1 451	23 007	95.0
宁 波 Ningbo	12 893	5 390	11 633	90.0
温 州 Wenzhou	5 727	411	4 974	87.0
嘉 兴 Jiaxing	8 059	475	7 976	99.0
绍 兴 Shaoxing	22 119	2	21 731	98.0
舟 山 Zhoushan	1 856	1 198	1 787	96.0
台 州 Taizhou	3 752	338	3 377	90.0
福 建 Fujian	**105 682**	**73 706**	**105 076**	**99.0**
福 州 Fuzhou	2 779	376	2 432	88.0
厦 门 Xiamen	3 776	155	3 755	99.0
莆 田 Putian	1 571	486	1 498	95.0
泉 州 Quanzhou	22 612	1 773	22 574	100.0
漳 州 Zhangzhou	73 521	70 909	73 433	100.0
宁 德 Ningde	1 423	7	1 384	97.0
山 东 Shandong	**38 265**	**5 608**	**38 135**	**100.0**
青 岛 Qingdao	3 831	232	3 822	100.0
东 营 Dongying	10 079	0	10 079	100.0
烟 台 Yantai	6 529	2 172	6 529	100.0
潍 坊 Weifang	7 095	0	7 065	100.0
威 海 Weihai	1 300	255	1 300	100.0
日 照 Rizhao	5 414	2 949	5 414	100.0
滨 州 Binzhou	4 017	0	3 926	98.0

8-7 续表2 continued

地 区 Region	工业废水排放总量 Total Volume of Industrial Waste Water Discharged	直接排入海的 Discharged Directly to Sea	工业废水中 Industrial Waste Water 符合排放标准的 Up-to-the-Standard	达标率（%） Up-to-the-Standard Rate (%)
广 东 Guangdong	**98 990**	**6 477**	**92 120**	**93.0**
广 州 Guangzhou	11 207	1 992	10 806	96.0
深 圳 Shenzhen	8 073	606	7 776	96.0
珠 海 Zhuhai	4 003	61	3 919	98.0
汕 头 Shantou	4 942	95	4 465	90.0
江 门 Jiangmen	8 504	345	8 116	95.0
湛 江 Zhanjiang	4 345	159	3 669	84.0
茂 名 Maoming	5 008	914	4 103	82.0
惠 州 Huizhou	1 971	0	1 936	98.0
汕 尾 Shanwei	4 161	3	1 960	47.0
阳 江 Yangjiang	950	19	759	80.0
东 莞 Dongguan	29 962	2 222	29 310	98.0
中 山 Zhongshan	12 629	41	12 396	98.0
潮 州 Chaozhou	1 285	0	1 070	83.0
揭 阳 Jieyang	1 950	20	1 835	94.0
广 西 Guangxi	**5 721**	**1 596**	**5 265**	**92.0**
北 海 Beihai	914	76	826	90.0
防城港 Fangchenggang	2 565	583	2 299	90.0
钦 州 Qinzhou	2 242	937	2 140	95.0
海 南 Hainan	**479**	**0**	**479**	**100.0**
海 口 Haikou	454	0	454	100.0
三 亚 Sanya	25	0	25	100.0

注：表中数据为沿海地带合计数（表8-10、表8-13同）。

Note: The data in the table are the total numbers for the coastal regions(The same as in the Tables 8-10,8-13).

8-8 沿海地区工业固体废物排放、处理及综合利用情况
Discharge, Treatment and Utilization of Industrial Solid Wastes by Coastal Regions

地　区 Region		工业固体废物排放量（吨） Volume of Industrial Solid Wastes Discharged (ton)	工业固体废物处置量（万吨） Volume of Industrial Solid Wastes Treated (10 000 tons)	工业固体废物综合利用量（万吨） Volume of Industrial Solid Wastes Muti-Utilized (10 000 tons)
总 计 Total		**645 223**	**161 577 948**	**656 636 883**
环渤海经济区 Round-the-Bohai Sea Economic Zone	**合　计 Total**	**332 205**	**130 705 839**	**392 586 843**
	辽　宁 Liaoning	27 510	72 610 978	82 406 816
	河　北 Hebei	304 551	52 598 783	156 933 243
	天　津 Tianjin	0	256 723	14 983 044
	山　东 Shandong	144	5 239 355	138 263 740
长江三角洲经济区 Yangzi River Delta Economic Zone	**合　计 Total**	**7 866**	**4 436 257**	**126 330 000**
	江　苏 Jiangsu	0	1 020 086	78 622 471
	上　海 Shanghai	42	856 579	21 715 950
	浙　江 Zhejiang	7 824	2 559 592	35 858 543
海峡西岸经济区 Economic Zone on West Side of the Straits	**合　计 Total**	**24 334**	**8 744 709**	**54 258 121**
	福　建 Fujian	24 334	8 744 709	54 258 121
珠江三角洲经济区 Zhujiang River Delta Economic Zone	**合　计 Total**	**159 839**	**3 134 850**	**43 215 626**
	广　东 Guangdong	159 839	3 134 850	43 215 626
环北部湾经济区 Round-the-Beibu Gulf Economic Zone	**合　计 Total**	**120 979**	**14 556 293**	**40 246 293**
	广　西 Guangxi	120 979	14 344 218	38 566 833
	海　南 Hainan	0	212 075	1 679 460

8-9 沿海城市工业固体废物排放、处理、综合利用情况
Discharge, Treatment and Utilization of Industrial Solid Wastes by Coastal Cities

单位：吨 (ton)

沿海城市 Coastal City	工业固体废物排放量 Volume of Industrial Solid Wastes Discharged	工业固体废物处置量 Volume of Industrial Solid Wastes Treated	工业固体废物综合利用量 Volume of Industrial Solid Wastes Muti-Utilized
天　津　Tianjin	0	256 723	14 983 044
唐　山　Tangshan	254 840	14 028 510	73 631 124
秦皇岛　Qinhuangdao	0	1 202 464	6 865 884
沧　州　Cangzhou	0	9 441	2 177 894
大　连　Dalian	7 798	185 368	3 572 526
丹　东　Dandong	0	1 185 971	1 493 755
锦　州　Jinzhou	155	1 879 376	1 783 054
营　口　Yingkou	0	4 263	4 742 875
盘　锦　Panjin	2	57 282	939 811
葫芦岛　Huludao	19 348	575 192	2 152 313
上　海　Shanghai	42	856 579	21 715 950
南　通　Nantong	0	40 256	3 033 035
连云港　Lianyungang	0	2 424	2 282 017
盐　城　Yancheng	0	47 047	1 645 281

8-9 续表1 continued

沿海城市 Coastal City	工业固体废物排放量 Volume of Industrial Solid Wastes Discharged	工业固体废物处置量 Volume of Industrial Solid Wastes Treated	工业固体废物综合利用量 Volume of Industrial Solid Wastes Muti-Utilized
杭 州 Hangzhou	2 651	286 037	6 061 942
宁 波 Ningbo	500	1 086 258	8 118 828
温 州 Wenzhou	1 892	109 063	1 836 189
嘉 兴 Jiaxing	0	75 174	3 578 705
绍 兴 Shaoxing	0	614 559	2 711 203
舟 山 Zhoushan	310	4 318	627 464
台 州 Taizhou	1 231	137 775	2 250 708
福 州 Fuzhou	6 100	122 386	4 011 622
厦 门 Xiamen	0	42 737	1 195 049
莆 田 Putian	0	545	372 180
泉 州 Quanzhou	5 807	220 910	4 711 087
漳 州 Zhangzhou	337	12 770	1 905 848
宁 德 Ningde	1 694	98 712	671 899
青 岛 Qingdao	0	92 433	9 036 895
东 营 Dongying	0	131 875	1 833 860
烟 台 Yantai	0	2 384 641	16 293 701
潍 坊 Weifang	0	430 899	6 233 017
威 海 Weihai	0	179 874	1 908 324
日 照 Rizhao	0	375	7 242 539
滨 州 Binzhou	100	31	7 028 014

8-9 续表2 continued

沿海城市 Coastal City	工业固体废物排放量 Volume of Industrial Solid Wastes Discharged	工业固体废物处置量 Volume of Industrial Solid Wastes Treated	工业固体废物综合利用量 Volume of Industrial Solid Wastes Muti-Utilized
广　州　Guangzhou	260	424 751	5 979 499
深　圳　Shenzhen	500	137 159	1 256 830
珠　海　Zhuhai	3 500	35 085	2 585 059
汕　头　Shantou	107	8 578	699 975
江　门　Jiangmen	4 006	29 841	2 033 524
湛　江　Zhanjiang	9 484	192 380	2 480 372
茂　名　Maoming	2 401	59 417	1 437 806
惠　州　Huizhou	0	54 160	218 741
汕　尾　Shanwei	20 115	29 263	342 615
阳　江　Yangjiang	6 621	87	781 897
东　莞　Dongguan	7 182	153 143	3 023 933
中　山　Zhongshan	4 895	139 236	705 787
潮　州　Chaozhou	1 300	377	609 775
揭　阳　Jieyang	145	1 861	547 790
北　海　Beihai	1 794	647 429	728 236
防城港　Fangchenggang	1 800	0	1 478 779
钦　州　Qinzhou	6 100	13 917	717 862
海　口　Haikou	0	562	42 224
三　亚　Sanya	0	23	2 711

8-10 沿海地带工业固体废物排放、处理、综合利用情况
Discharge,Treatment and Utilization of Industrial Solid Wastes by Coastal Counties

单位：吨 (ton)

地 区 Region	工业固体废物排放量 Volume of Industrial Solid Wastes Discharged	工业固体废物处置量 Volume of Industrial Solid Wastes Treated	工业固体废物综合利用量 Volume of Industrial Solid Wastes Muti-Utilized
天 津 Tianjin	**0**	**189 621**	**5 010 622**
河 北 Hebei	**0**	**29 869**	**15 526 288**
唐 山 Tangshan	0	18 724	10 458 256
秦皇岛 Qinhuangdao	0	6 944	3 505 147
沧 州 Cangzhou	0	4 201	1 562 885
辽 宁 Liaoning	**7 921**	**743 466**	**9 914 468**
大 连 Dalian	7 798	157 761	3 317 649
丹 东 Dandong	0	50	474 126
锦 州 Jinzhou	0	6 219	259 035
营 口 Yingkou	0	4 244	4 488 956
盘 锦 Panjin	0	0	176 383
葫芦岛 Huludao	123	575 192	1 198 319
上 海 Shanghai	**40**	**670 444**	**18 759 860**
江 苏 Jiangsu	**0**	**32 296**	**3 612 817**
南 通 Nantong	0	14 946	635 437
连云港 Lianyungang	0	2 038	2 227 243
盐 城 Yancheng	0	15 312	750 137

8-10 续表1 continued

地　区 Region	工业固体废物排放量 Volume of Industrial Solid Wastes Discharged	工业固体废物处置量 Volume of Industrial Solid Wastes Treated	工业固体废物综合利用量 Volume of Industrial Solid Wastes Muti-Utilized
浙　江 Zhejiang	**2 873**	**1 506 568**	**17 008 967**
杭　州 Hangzhou	0	89 901	2 445 387
宁　波 Ningbo	300	1 033 554	6 458 263
温　州 Wenzhou	1 032	71 658	1 729 103
嘉　兴 Jiaxing	0	43 285	2 185 427
绍　兴 Shaoxing	0	130 859	1 511 954
舟　山 Zhoushan	310	4 318	627 464
台　州 Taizhou	1 231	132 993	2 051 369
福　建 Fujian	**11 390**	**128 295**	**9 353 898**
福　州 Fuzhou	6 100	1 455	3 590 925
厦　门 Xiamen	0	42 737	1 195 049
莆　田 Putian	0	545	372 180
泉　州 Quanzhou	4 253	42 434	2 218 228
漳　州 Zhangzhou	0	12 017	1 383 799
宁　德 Ningde	1 037	29 107	593 717
山　东 Shandong	**0**	**2 673 922**	**32 304 016**
青　岛 Qingdao	0	9 356	2 536 030
东　营 Dongying	0	131 875	1 833 860
烟　台 Yantai	0	2 320 482	15 765 122
潍　坊 Weifang	0	36 031	1 763 812
威　海 Weihai	0	175 860	1 344 976
日　照 Rizhao	0	287	6 790 054
滨　州 Binzhou	0	31	2 270 162

8-10 续表2 continued

地　区 Region	工业固体废物排放量 Volume of Industrial Solid Wastes Discharged	工业固体废物处置量 Volume of Industrial Solid Wastes Treated	工业固体废物综合利用量 Volume of Industrial Solid Wastes Muti-Utilized
广　东　Guangdong	**50 470**	**945 533**	**16 064 569**
广　州　Guangzhou	80	218 934	4 059 940
深　圳　Shenzhen	500	137 159	1 256 830
珠　海　Zhuhai	2 000	21 844	275 980
汕　头　Shantou	107	8 574	699 975
江　门　Jiangmen	4 006	27 526	1 789 003
湛　江　Zhanjiang	7 181	177 008	2 388 652
茂　名　Maoming	2 401	25 633	869 879
惠　州　Huizhou	0	4 953	85 319
汕　尾　Shanwei	15 640	29 263	341 455
阳　江　Yangjiang	6 233	87	156 130
东　莞　Dongguan	7 182	153 143	3 023 933
中　山　Zhongshan	4 895	139 236	705 787
潮　州　Chaozhou	100	312	8 201
揭　阳　Jieyang	145	1 861	403 485
广　西　Guangxi	**2 569**	**661 346**	**1 451 024**
北　海　Beihai	769	647 429	297 904
防城港　Fangchenggang	1 800	0	701 080
钦　州　Qinzhou	0	13 917	452 040
海　南　Hainan	**0**	**583**	**23 753**
海　口　Haikou	0	560	21 042
三　亚　Sanya	0	23	2 711

8-11 沿海地区污染治理项目情况
Pollution Control Projects by Coastal Regions

单位：个 (number)

地 区 Region		当年安排施工项目 Arranged for Construction in the Year		当年竣工项目 Completed in the Year	
		治理废水 Treatment of Waste Water	治理固体废物 Treatment of Solid Wastes	治理废水 Treatment of Waste Water	治理固体废物 Treatment of Solid Wastes
总 计 Total		**2 224**	**178**	**1 976**	**159**
环渤海经济区 Round-the-Bohai Sea Economic Zone	**合 计 Total**	**564**	**63**	**512**	**61**
	辽 宁 Liaoning	68	8	59	7
	河 北 Hebei	108	0	103	0
	天 津 Tianjin	48	7	41	7
	山 东 Shandong	340	48	309	47
长江三角洲经济区 Yangzi River Delta Economic Zone	**合 计 Total**	**847**	**38**	**768**	**36**
	江 苏 Jiangsu	379	19	351	19
	上 海 Shanghai	80	5	69	5
	浙 江 Zhejiang	388	14	348	12
海峡西岸经济区 Economic Zone on West Side of the Straits	**合 计 Total**	**251**	**25**	**204**	**16**
	福 建 Fujian	251	25	204	16
珠江三角洲经济区 Zhujiang River Delta Economic Zone	**合 计 Total**	**410**	**38**	**361**	**34**
	广 东 Guangdong	410	38	361	34
环北部湾经济区 Round-the-Beibu Gulf Economic Zone	**合 计 Total**	**152**	**14**	**131**	**12**
	广 西 Guangxi	143	14	124	12
	海 南 Hainan	9	0	7	0

8-12 沿海城市污染治理项目情况
Pollution Control Projects by Coastal Cities

单位：个 (number)

沿海城市 Coastal City	当年安排施工项目 Arranged for Construction in the Year		当年竣工项目 Completed in the Year	
	治理废水 Treatment of Waste Water	治理固体废物 Treatment of Solid Wastes	治理废水 Treament of Waste Water	治理固体废物 Treatment of Solid Wastes
天　津　Tianjin	48	7	41	7
唐　山　Tangshan	13	0	12	0
秦皇岛　Qinhuangdao	19	0	19	0
沧　州　Cangzhou	5	0	5	0
大　连　Dalian	18	0	13	0
丹　东　Dandong	1	0	1	0
锦　州　Jinzhou	5	2	2	1
营　口　Yingkou	12	0	11	0
盘　锦　Panjin	1	1	1	1
葫芦岛　Huludao	7	1	7	1
上　海　Shanghai	80	5	69	5
南　通　Nantong	29	3	29	3
连云港　Lianyungang	8	1	8	1
盐　城　Yancheng	3	1	3	1

8-12 续表1 continued

沿海城市 Coastal City	当年安排施工项目 Arranged for Construction in the Year		当年竣工项目 Completed in the Year	
	治理废水 Treatment of Waste Water	治理固体废物 Treatment of Solid Wastes	治理废水 Treament of Waste Water	治理固体废物 Treatment of Solid Wastes
杭　州 Hangzhou	20	4	16	3
宁　波 Ningbo	23	1	20	1
温　州 Wenzhou	58	0	56	0
嘉　兴 Jiaxing	47	1	43	1
绍　兴 Shaoxing	51	4	46	4
舟　山 Zhoushan	6	0	6	0
台　州 Taizhou	40	2	33	1
福　州 Fuzhou	16	3	15	3
厦　门 Xiamen	23	1	21	1
莆　田 Putian	4	1	4	1
泉　州 Quanzhou	122	12	87	3
漳　州 Zhangzhou	9	0	10	0
宁　德 Ningde	0	0	0	0
青　岛 Qingdao	25	3	25	2
东　营 Dongying	28	3	22	3
烟　台 Yantai	14	37	13	37
潍　坊 Weifang	44	1	39	1
威　海 Weihai	14	1	10	1
日　照 Rizhao	17	0	17	0
滨　州 Binzhou	54	0	51	0

8-12 续表2 continued

沿海城市 Coastal City	当年安排施工项目 Arranged for Construction in the Year		当年竣工项目 Completed in the Year	
	治理废水 Treatment of Waste Water	治理固体废物 Treatment of Solid Wastes	治理废水 Treament of Waste Water	治理固体废物 Treatment of Solid Wastes
广 州 Guangzhou	24	1	18	1
深 圳 Shenzhen	89	0	66	0
珠 海 Zhuhai	11	0	11	0
汕 头 Shantou	31	0	29	0
江 门 Jiangmen	8	2	7	2
湛 江 Zhanjiang	21	0	21	0
茂 名 Maoming	1	0	1	0
惠 州 Huizhou	47	5	47	5
汕 尾 Shanwei				
阳 江 Yangjiang	9	0	7	0
东 莞 Dongguan	6	2	6	2
中 山 Zhongshan	8	2	7	1
潮 州 Chaozhou	4	1	4	0
揭 阳 Jieyang	3	0	3	0
北 海 Beihai	5	0	3	0
防城港 Fangchenggang	4	0	4	0
钦 州 Qinzhou	3	1	3	1
海 口 Haikou				
三 亚 Sanya				

8-13 沿海地带污染治理项目情况
Pollution Control Projects by Coastal Counties

单位：个 (number)

地 区 Region	当年安排施工项目 Arranged for Construction in the Year		当年竣工项目 Completed in the Year	
	治理废水 Treatment of Waste Water	治理固体废物 Treatment of Solid Wastes	治理废水 Treament of Waste Water	治理固体废物 Treatment of Solid Wastes
天 津 Tianjin	**16**	**1**	**16**	**1**
河 北 Hebei	**1**	**0**	**1**	**0**
唐 山 Tangshan	0	0	0	0
秦皇岛 Qinhuangdao	0	0	0	0
沧 州 Cangzhou	1	0	1	0
辽 宁 Liaoning	**30**	**3**	**24**	**2**
大 连 Dalian	16	0	11	0
丹 东 Dandong	0	0	0	0
锦 州 Jinzhou	1	2	0	1
营 口 Yingkou	5	0	5	0
盘 锦 Panjin	1	0	1	0
葫芦岛 Huludao	7	1	7	1
上 海 Shanghai	**60**	**5**	**52**	**5**
江 苏 Jiangsu	**22**	**1**	**22**	**1**
南 通 Nantong	11	0	11	0
连云港 Lianyungang	8	1	8	1
盐 城 Yancheng	3	0	3	0

8-13 续表1 continued

地 区 Region	当年安排施工项目 Arranged for Construction in the Year		当年竣工项目 Completed in the Year	
	治理废水 Treatment of Waste Water	治理固体废物 Treatment of Solid Wastes	治理废水 Treament of Waste Water	治理固体废物 Treatment of Solid Wastes
浙 江 Zhejiang	**162**	**5**	**143**	**5**
杭 州 Hangzhou	4	1	3	1
宁 波 Ningbo	14	0	11	0
温 州 Wenzhou	58	0	56	0
嘉 兴 Jiaxing	18	1	14	1
绍 兴 Shaoxing	33	3	28	3
舟 山 Zhoushan	6	0	6	0
台 州 Taizhou	29	0	25	0
福 建 Fujian	**144**	**13**	**107**	**4**
福 州 Fuzhou	9	1	9	1
厦 门 Xiamen	23	1	21	1
莆 田 Putian	4	1	4	1
泉 州 Quanzhou	105	10	70	1
漳 州 Zhangzhou	3	0	3	0
宁 德 Ningde	0	0	0	0
山 东 Shandong	**127**	**41**	**113**	**41**
青 岛 Qingdao	7	1	7	1
东 营 Dongying	28	3	22	3
烟 台 Yantai	14	37	13	37
潍 坊 Weifang	9	0	7	0
威 海 Weihai	8	0	6	0
日 照 Rizhao	11	0	11	0
滨 州 Binzhou	50	0	47	0

8-13 续表2 continued

地 区 Region	当年安排施工项目 Arranged for Construction in the Year		当年竣工项目 Completed in the Year	
	治理废水 Treatment of Waste Water	治理固体废物 Treatment of Solid Wastes	治理废水 Treament of Waste Water	治理固体废物 Treatment of Solid Wastes
广 东 Guangdong	**202**	**5**	**171**	**4**
广 州 Guangzhou	11	0	7	0
深 圳 Shenzhen	89	0	66	0
珠 海 Zhuhai	9	0	9	0
汕 头 Shantou	31	0	29	0
江 门 Jiangmen	7	1	6	1
湛 江 Zhanjiang	21	0	21	0
茂 名 Maoming	1	0	1	0
惠 州 Huizhou	10	0	10	0
汕 尾 Shanwei	0	0	0	0
阳 江 Yangjiang	3	0	3	0
东 莞 Dongguan	6	2	6	2
中 山 Zhongshan	8	2	7	1
潮 州 Chaozhou	4	0	4	0
揭 阳 Jieyang	2	0	2	0
广 西 Guangxi	**7**	**0**	**5**	**0**
北 海 Beihai	4	0	2	0
防城港 Fangchenggang	1	0	1	0
钦 州 Qinzhou	2	0	2	0
海 南 Hainan	**0**	**0**	**0**	**0**
海 口 Haikou	0	0	0	0
三 亚 Sanya	0	0	0	0

8-14 沿海区域海洋类型自然保护区建设情况
Construction of Marine-Type Nature Reserves

地　区 Region	保护区数量（个） Number of Nature Reserves (number)		按保护级别分（个） By Level of Protection (number)		按保护类型分（个） By Type of Protection (number)			保护区面积（平方公里） Area(km^2)
	已建 Already Established	新建 Newly Established	国家级 National	地方级 Provincial	海洋和海岸生态系统 Marine and Coastal Ecosystem	海洋自然历史遗迹 Marine Natural and Historical Relics	海洋生物多样性 Marine Biodiversity	
合　计 Total	**157**		**32**	**125**	**93**	**15**	**49**	**29 460.77**
环渤海经济区 Round-the-Bohai Sea Economic Zone	31		11	20	19	5	7	15 682.73
长江三角洲经济区 Yangzi River Delta Economic Zone	11		5	6	9		2	5 679.22
海峡西岸经济区 Economic Zone on West Side of the Straits	15		3	12	9	2	4	1 592.41
珠江三角洲经济区 Zhujiang River Delta Economic Zone	68		6	62	33	3	32	4 340.04
环北部湾经济区 Round-the-Beibu Gulf Economic Zone	32		7	25	23	5	4	2 166.37

8-15 沿海地区海洋类型自然保护区建设情况
Construction of Marine-Type Nature Reserves

地 区 Region	保护区数量（个） Number of Nature Reserves (number)		按保护级别（个） By Level of Protection (number)		按保护类型分（个） By Type of Protection (number)			保护区面积（平方公里） Area(km^2)
	已建 Already Established	新建 Newly Established	国家级 National	地方级 Provincial	海洋和海岸生态系统 Marine and Coastal Ecosystem	海洋自然历史遗迹 Marine Natural and Historical Relics	海洋生物多样性 Marine Biodiversity	
合 计 Total	**157**		**32**	**125**	**93**	**15**	**49**	**29 460.77**
天 津 Tianjin	1		1			1		359.13
河 北 Hebei	4		1	3	3	1		403.07
辽 宁 Liaoning	14		5	9	7	3	4	11 714.10
上 海 Shanghai	4		2	2	3		1	982.75
江 苏 Jiangsu	4		2	2	3		1	3 350.76
浙 江 Zhejiang	3		1	2	3			1 345.71
福 建 Fujian	15		3	12	9	2	4	1 592.41
山 东 Shandong	12		4	8	9		3	3 206.43
广 东 Guangdong	68		6	62	33	3	32	4 340.04
广 西 Guangxi	6		3	3	5	1		806.78
海 南 Hainan	26		4	22	18	4	4	1 359.59

注:全国海洋特别保护区43个，辽宁1个、山东10个、江苏2个、浙江4个，福建24个，广东1个，海南1个。

Note: There are 43 national special marine protected areas, i.e,1 in Liaoning, 10 in Shandong, 2 in Jiangsu, 4 in Zhejiang, 24 in Fujian, 1 in Guangdong, 1 in Hainan.

8-16 全国海洋生态监控区基本情况
Basic Condition of the Marine Ecological Monitoring Areas Throughout the Country

生态监控区 Ecological Monitoring Area	所在地 Location	面积（平方公里） Area (km^2)	主要生态系统类型 Major Types of Ecosystem	健康状况 Health Condition
双台子河口	辽宁省	3 000	河 口	亚健康
Shuangtaizi Estuary	Liaoning Province		Estuary	Subhealthy
锦州湾*	辽宁省	650	海 湾	不健康
Jinzhou Bay	Liaoning Province		Bay	unhealthy
滦河口-北戴河	河北省	900	河 口	亚健康
Luanhe Mouth-Beidaihe	Hebei Province		Estuary	Subhealthy
渤海湾	天津市	3 000	海 湾	不健康
Bohai Bay	Tianjin Municipality		Bay	unhealthy
莱州湾	山东省	3 770	海 湾	不健康
Laizhou Bay	Shandong Province		Bay	unhealthy
黄河口	山东省	2 600	河 口	亚健康
Yellow River Mouth	Shandong Province		Estuary	Subhealthy
苏北浅滩	江苏省	3 090	湿 地	亚健康
North Jiangsu Bank	Jiangsu Province		Wetland	Subhealthy
长江口	上海市	13 668	河 口	亚健康
Changjiang River Mouth	Shanghai Municipality		Estuary	Subhealthy
杭州湾	上海市 浙江省	5 000	海 湾	不健康
Hangzhou Bay	Shanghai Municipality Zhejiang Province		Bay	unhealthy
乐清湾	浙江省	464	海 湾	亚健康
Leqing Bay	Zhejiang Province		Bay	Subhealthy

8-16 续表 continued

生态监控区 Ecological Monitoring Area	所在地 Location	面积(平方公里) Area (km^2)	主要生态系统类型 Major Types of Ecosystem	健康状况 Health Condition
闽东沿岸	福建省	5 063	海 湾	亚健康
Constal East Fujian	Fujian Province		Bay	Subhealthy
大亚湾	广东省	1 200	海 湾	亚健康
Daya Bay	Guangdong Province		Bay	Subhealthy
珠江口	广东省	3 980	河 口	不健康
Zhujiang River Mouth	Guangdong Province		Estuary	unhealthy
雷州半岛 西南沿岸	广东省	1 150	珊瑚礁	亚健康
Southwest Coast of Leizhou Peninsula	Guangdong Province		Coral Reef	Subhealthy
广西北海	广西壮族自治区	120	珊瑚礁 红树林 海草床	健 康 健 康 亚健康
Beihai,Guangxi	Guangxi Zhuang Nationality Autonomous Region		Coral Reef Mangroves Seagrass Bed	Healthy Healthy Subhealthy
北仑河口*	广西壮族自治区	150	红树林	健 康
Beilun River Mouth	Guangxi Zhuang Nationality Autonomous Region		Mangroves	Healthy
海南东海岸	海南省	3 750	珊瑚礁 海草床	健 康 健 康
East Coast of Hainan	Hainan Province		Coral Reef Seagrass Bed	Healthy Healthy
西沙珊瑚礁*	海南省	400	珊瑚礁	亚健康
Xisha Coral Reef	Hainan Province		Coral Reef	Subhealthy

注：*2005年新增生态监控区，变化趋势指四年。
Note: *The newly added ecological monitoring areas with four-year variation trends.

8-17 沿海地区风暴潮灾害情况
Survey of Storm Surges by Coastal Regions

受灾地区 Disaster Area	受灾人口（万人） Disaster-stricken Population (10 000 persons)	农作物受灾（千公顷） Disaster-Affected crops (1 000 hm^2)	损失船只（艘） Lost Boat (Number)	海水养殖受灾面积（千公顷） Affected Area of Mariculture (1 000 hm^2)
合 计 Total	**872.12**	**436.32**	**3 047**	**99.85**
天 津 Tianjin	0.00	0.00	2	
河 北 Hebei	5.00	0.00	20	7.00
辽 宁 Liaoning	0.00	0.00	0	0.20
上 海 Shanghai	0.00	0.00	0	0.00
江 苏 Jiangsu	0.00	0.00	186	20.26
浙 江 Zhejiang	0.00	0.00	920	42.21
福 建 Fujian	165.00	66.06	1 152	7.46
山 东 Shandong	6.50	0.00	24	2.27
广 东 Guangdong	361.24	274.05	0	15.90
广 西 Guangxi	8.45	4.05	3	0.00
海 南 Hainan	325.93	92.16	740	4.55

8-17 续表 continued

受灾地区 Disaster Area	损毁海岸工程（公里） Destroyed Coastal Engineering Architectures(km)	死亡人数*（人） Death Toll (person)	直接经济损失（亿元） Direct Economic Loss (100 million yuan)
合 计 Total	**267.81**	**57**	**84.97**
天 津 Tianjin	16.40	9	2.49
河 北 Hebei	23.50	0	0.70
辽 宁 Liaoning	2.80	0	0.74
上 海 Shanghai	0.00	0	0.00
江 苏 Jiangsu	40.00	2	0.97
浙 江 Zhejiang	13.82	1	11.85
福 建 Fujian	27.12	4	19.83
山 东 Shandong	5.40	0	3.01
广 东 Guangdong	136.97	21	38.99
广 西 Guangxi	0.42	8	0.14
海 南 Hainan	1.38	12	6.25

注：*包括失踪人数。

Notes:*Includes lost people.

8-18 沿海地区赤潮灾害情况*

Survey of Red Tide by Coastal Regions

时间 Date	影响区域 Area		最大面积（平方公里） Disaster Area(km^2)
合　计　Total			**12 940**
其中：Including:			
4月9日 Apr.9	长江口海域	Changjiang River Mouth Waters	100
4月28日 Apr.28	台州外侧海域	Sea Area on the Outer Flank of Taizhou	700
5月2日至5月7日 May.2-May.7	渔山列岛—台州列岛海域	Yushan Archipelago-Taizhou Archipelago Waters	1 330
5月7日至5月12日 May.7-May.12	山东日照附近海域	Sea Area Near Rizhao, Shandong	580
5月7日至5月12日 May.7-May.12	温州苍南大渔湾海域	Dayuwan Area, Cangnan, Wenzhou	200
5月10日 May.10	舟山北部海域	Northern Waters of Zhoushan	360
5月16日至5月18日 May.16-May.18	台湾海峡东碇岛以东海域	Sea Area to the East of Dongding Island in the Taiwan Straits	200
5月19日至5月30日 May.19-May.30	长江口外、舟山北部海域	Sea Area off the Changjiang River Mouth and Northern Zhoushan	1 500
5月26日至6月1日 May.26-Jun.1	山东海阳至乳山附近海域	Sea Area from Haiyang, Shandong to Rushan	550
5月26日至6月1日 May.26-Jun.1	河北昌黎新开口附近海域	Sea Area Adjacent to Xinkaikou, Changli, Hebei	460
5月31日至6月13日 May.31-Jun.13	渤海湾附近海域	Sea Area Near the Bohai Bay	4 460

8-18 续表 continued

时间 Date	影响区域 Area		最大面积（平方公里） Disaster Area(km^2)
6月4日 Jun.14	舟山北部海域	Northern Waters of Zhoushan	400
6月11日 Jun.11	江苏南通外海海域	Off-sea Area of Nantong, Jiangsu	350
6月17日至6月22日 Jun.17-Jun.22	舟山朱家尖岛以东海域	Sea Areas East of Zhujiajian Island, Zhoushan	310
6月17日至6月22日 Jun.17-Jun.22	嵊山西南海域	Sea Area Southwest of Shengshan	230
7月18日至7月23日 Jul.18-Jul.23	连云港市东西连岛及拦海大堤东北海域	Sea Area Northeast of Lianyungang City's East-west Tied Island and Dam Across the Sea	210
7月27日 Jul.27	舟山岛东部海域	Eastern Waters of Zhoushan Island	300
8月1日至8月3日 Aug.1-Aug.3	天津港航道以北至汉沽近岸海域	Nearshore Sea Area from North of the Tianjin Port Channel to Hangu	300
8月4日至8月5日 Aug.4-Aug.5	舟山朱家尖东部海域	Eastern Waters of Zhujiajian, Zhoushan	120
10月27日至11月9日 Oct.27-Num.9	珠海市淇澳岛附近海域	Sea Area Near Qi'ao Island, Zhuhai	280

注：*为我国近海面积100平方公里以上的赤潮。

Note: *The data is for the red tide over 100 square kitometers in China coastal area.

主要统计指标解释

1. 工业废水排放量 指经过企业厂区所有排放口排到企业外部的工业废水量。包括生产废水、外排的直接冷却水、超标排放的矿井地下水和与工业废水混排的厂区生活污水,不包括外排的间接冷却水(清污不分流的间接冷却水应计算在内)。

2. 直接排入海的工业废水量 指经企业位于海边的排放口，直接排入海中的废水量。直接排入是指废水经过工厂的排污口直接排入海，而未经过城市下水道或其他中间体，也不受其他水体的影响。

3. 工业废水处理量 指报告期内各种水治理设施实际处理的工业废水量,包括处理后外排的和处理后回用的工业废水量，虽经处理但未达到国家或地方排放标准的废水量也应计算在内。计算时，如遇车间和厂排放口均有治理设施，并对同一废水分级处理时，不应重复计算工业废水处理量。

4. 工业废水处理率 指工业废水处理量占需要处理的工业废水量的百分率。其计算公式是:

工业废水处理率 =（工业废水处理量 / 需处理的工业废水量）×100%

式中，需处理工业废水量 =工业废水排放量+工业废水处理回用量-(工业废水排放达标量-工业废水处理排放达标量)

5. 工业废水排放达标量 指各项指标都达到国家或地方排放标准的外排工业废水量,包括经过处理后外排达标的和未经处理外排达标的两部分。国家排放标准见 GB 8978-88。

6. 工业废水处理排放达标量 指经过各种水治理设施处理后达到国家或地方排放标准的废水排放量。

7. 工业固体废物处置量 指将固体废物焚烧或者最终置于符合环境保护规定要求的场所并不再回取的工业固体废物量(包括当年处置往年的工业固体废物累计贮存量)。

8. 工业固体废物排放量 指将所产生的固体废物排到固体废物污染防治设施、场所以外的量。不包括矿山开采的剥离废石和掘进废石(煤矸石和呈酸性或碱性的废石除外)。

9. 当年开工污染治理项目数 指报告期内由国家、部门、地方或企业单位安排开工的，并以治理废水、废气、固体废物、噪声和其他(如电磁波、恶臭等)环境污染的环境治理工程的总数。不包括“三同时”项目。

10. 当年竣工项目数 指报告期内竣工投入运行的治理废水、废气、固体废物、噪声及其他污染的环境工程项目的总数。

Explanatory Notes on Main Statistical Indicators

1. Volume of Industrial Waste Water Discharged refers to the quantity of industrial waste water discharged externally through all the outlets in the factory area of the enterprise, including the waste water from production, externally discharged direct cooling water, mine-shaft groundwater discharged exceeding the set standard and the domestic sewage of the factory area discharged together with the industrial waste water, but not including the indirect cooling water discharge externally.

2. Volume of Industrial Waste Water discharged Directly to Sea refers to the quantity of waste water directly discharged into the sea through the outlets of the enterprise by the sea. Direct discharge means the direct discharge into the sea of waste water through the outlets of the factory, which is not discharged via the urban sewers or other intermediates and is not affected by other water bodies.

3. Volume of Industrial Waste Water Treated refers to the industrial waste water volume actually treated by various water treatment facilities in the period covered by the report, including the quantity of the industrial waste water discharged and reused after treatment. The amount of the waste water which is not up to the state or local standard of discharge upon treatment should be included. If the workshops and outlets of the factory are provided with treatment facilities and carry out graded treatment of the same waste water, the processed volume of industrial waste water cannot be calculated repeatedly.

4. Rate of Industrial Waste Water Treatment refers to the percentage of the processed volume of industrial waste water in the industrial waste water volume that needs to be treated. The calculation formula is:

Rate of Industrial Waste Water Treatment = (Processed Volume of Industrial Waste Water/Processed Volume of Industrial Waste Water Required) × 100%.

Where: Processed Volume of Industrial Waste Water Required = (Processed Volume of Industrial Waste Water + Reused Volume of Industrial Waste Water After Treatment−Up-to-Standard Volume of Industrial Waste Water for Discharge−Up-to-Standard Volume of Industrial Waste Water for Discharge after Treatment).

5. Volume of Up-to-Standard Industrial Waste Water Discharged refers to the externally discharged volume of industrial waste water with all its indexes up to the state or local standard for discharge, including the volume of industrial waste water up to the standard for discharge after being treated or that without being treated. See GB8978−88 for the state's discharge standard.

6. Volume of Up-to-Standard Discharged Industrial Waste Water After Treatment refers to the discharged volume of waste water that is up to the state or local standard for discharge upon treatment by various water treatment facilities.

7. Volume of Industrial Solid Wastes Discharged refers to the amount of solid wastes discharged out of the facilities and sites for the solid wastes pollution prevention and control, not including the stripped and tunneled waste ores in the excavation of mines (other than gangues and acid or alkaline waste ores).

8. Volume of Industrial Solid Wastes Treated refers to the volume of industrial solid wastes which are to be burned or finally placed at the sites in keeping with the requirement of environmental protection and will not be recovered (including the accumulated amount of storage in former years of industrial solid radioactive matter disposed of in the year).

9. Number of Pollution Treatment Projects Started in the Current Year refers to the total number of environmental pollution control projects for controlling waste water, waste gas, solid wastes, noise and other environmental pollutions (such as electromagnetic wave, offensive odor) started by the state, governmental departments, local governments or enterprises in the period covered by the report mainly for the purpose of pollution control and multipurpose utilization of "three wastes".

10. Number of Pollution Treatment Projects Completed in the Current Year refers to the total number of environmental engineering projects for controlling waste water, waste gas, solid wastes, noise and other pollutions which are completed and put into operation in the period covered by the report.

9

海洋行政管理及公益服务

Marine Administration and Public-Good Service

9-1 海域使用管理情况
Sea Area Use Management

地 区 Region	发放海域使用权证书（本） Certificates of Right of Sea Area Use Issued (number)	确权海域面积（公顷） Area of Waters with Established Rights (hm^2)	海域使用金（万元） Charge for Sea Area Utilization (10 000 yuan)
全国总计 National Total	**5 327**	**178 366.86**	**786 251.54**
天 津 Tianjin	57	5 125.54	213 304.71
河 北 Hebei	88	3 755.35	14 539.50
辽 宁 Liaoning	758	77 064.91	101 334.64
上 海 Shanghai	8	793.03	1 992.34
江 苏 Jiangsu	218	31 045.06	13 628.39
浙 江 Zhejiang	355	11 873.79	58 037.74
福 建 Fujian	2 162	8 912.57	119 739.16
山 东 Shandong	374	25 623.51	121 719.81
广 东 Guangdong	491	10 104.60	63 824.77
广 西 Guangxi	169	2 323.90	51 579.88
海 南 Hainan	646	1 427.29	17 352.39
其 他* other	1	317.31	9 198.21

注：*为沿海省（自治区、直辖市）管理海域以外。

Note: * Sea areas outside the control of coastal provinces of autonomous resgions and municipalities directly under the Central Government

9-2 海洋倾废管理情况
Management on the Ocean Dumping of Wastes

海　区 Sea Area	签发疏浚物海洋倾倒许可证（份） Permit Issued for Ocean Dumping of Dredged Materials (number)	新选划倾倒区数（个） Number of Newly Designated Dumping Zones (number)
合　计 Total	**104**	**12**
渤黄海 Bohai and Huangai Sea	45	5
东　海 East China Sea	7	4
南　海 South China Sea	52	3

9-3 海洋执法检查情况
Marine Law Enforcement Inspection

执法检查项目分类 Classification of Items Subject to Low Enforcement Inspection	检查项目（个） Number of Items Subject to Inspection (number)	检查次数（次） Number of Times of Inspection (number)	发现违法行为（起） Illegal Acts Found (case)
合　计 Total	**28 611**	**67 499**	**1 817**
渔业用海 Sea use for fishery	20 334	37 632	884
交通运输用海 Sea use for transportation and communications	1 852	6 107	122
工矿用海 Sea use for industry and mining	1 891	5 475	158
旅游娱乐用海 Sea use for tourism and recreation	408	2 548	36
海底工程用海 Sea use for undersea engineering	105	566	4
排污倾倒用海 Sea use for sewage discharge and dumping	439	2 072	26
围海造地用海 Sea use for reclaiming land from sea	2 823	10 843	371
特殊用海 Special sea uses	125	431	18
其他用海 Other sea uses	634	1 825	198

9-4 沿海地区海滨观测台站分布概况
Distribution of Coastal Observation Stations by Coastal Regions

单位：个 (number)

地 区 Region	合 计 Total	海洋站 Marine Station	验潮站① Tide Station	气象台站② Meteorological Station	地震台站 Seismic Station	雷达站 Radar Station
合 计 Total	**768**	**74**	**235**	**102**	**249**	**108**
天 津 Tianjin	**12**	1	0	2	6	3
河 北 Hebei	**33**	4		4	20	5
辽 宁 Liaoning	**62**	7	2	13	30	10
上 海 Shanghai	**90**	4	63	1	10	12
江 苏 Jiangsu	**63**	4	19	19	19	2
浙 江 Zhejiang	**94**	9	35	14	20	16
福 建 Fujian	**86**	8	11	27	28	12
山 东 Shandong	**93**	8	21	8	41	15
广 东 Guangdong	**173**	15	76	7	53	22
广 西 Guangxi	**21**	5	5	3	4	4
海 南 Hainan	**40**	9	3	3	18	7

注： ①潮流量观测站35处，潮水位观测站200处。
②中国气象局所属的气象台站仅指沿海气象台站中有海洋气象观测、预报及服务等业务的台站，合计中含有中央气象台1个。

Notes: ①There are 35 tidal current observation stations and 200 tidal level observation stations.
② The meteorological observatories and stations under the China Meteorological Administration only refer to those with marine meteorotogical observation, forecast and service among the coastal meteorological observatories and the total includes the central meteorological station.

9-5 海洋预报服务概况
Marine Forecast Service

单位：次 (time)

预报项目 Item	数值预报 Numerical Forecast			
	预报服务次数 Frequency	发布次数 Frequency of Release		
		广播电视 Radio and TV	因特网 Internet	纸 质 Paper Media
合 计 Total	**369 061 242**	**2 555**	**18 777**	**369 054 107**
海 浪 Sea Wave	3 736	730	5 030	836
海 温 Sea Surface Temperature	177 675	1 095	5 030	175 930
潮 汐 Tide	368 176 206	730	2 166	368 174 826
海 流 Sea Current	176 610		4 315	175 215
海平面 Sea Level				
盐 度 Salinity	175 200		1 460	175 200
赤 潮 Red Tide	26		3	23
海水浴场 Bathing Beach	212		212	212
海 冰 Sea Ice	81		81	
风暴潮 Storm Surge	606		192	354
气 象 Metorology				
厄尔尼诺 EL Niño	12		12	720
专 项 Special Item	458		276	306
其 他 Others	350 420			350 485

9-5 续表 continued

预报项目 Item	统计预报 Statistical Forecast			
	预报服务次数 Frequency	发布次数 Frequency of Release		
		广播电视 Radio and TV	因特网 Internet	纸 质 Paper Media
合 计 Total	**321 009 053**	**70 306**	**320 830 328**	**150 376**
海 浪 Sea Wave	46 575	24 215	20 287	13 824
海 温 Sea Surface Temperature	34 408	20 719	14 473	10 048
潮 汐 Tide	33 324	20 709	13 676	9 561
海 流 Sea Current	320 859 820		320 765 650	94 900
海平面 Sea Level	110		110	
盐 度 Salinity	54			78
赤 潮 Red Tide	513		189	540
海水浴场 Bathing Beach	4 041	2 680	2 769	1 400
海 冰 Sea Ice	491	68	172	150
风暴潮 Storm Surge	2 358	672	273	1 981
气 象 Metorology	8 594	72	4 339	6 895
厄尔尼诺 EL Niño	917		13	1 629
专 项 Special Item	16 669	1 171	8 207	8 311
其 他 Others	1 179		170	1 059

注：本表只包含国家海洋局资料。

Note: This table only contains the data from the State Oceanic Administration.

9-6 海洋监测情况
Marine Monitoring Statistics

项目 Item	台站监测 Station Monitoring	断面监测 Sectional Monitoring	浮标监测 Buoy Monitoring	船舶测报 Ship measuring and reporting	其他监测 Other Monitoring
测站（点）数（个） Number of Station (unit)	175	153	39	228	1 458
实际获得数据量（个） Quantity of Data Actually Obtained (unit)	77 731 448	130 679	1 730 882	6 622 194	6 594 001

9-7 海洋调查概况
Marine Survey Statistics

调查名称 Name	站点数（个） Number of Stations (unit)	船舶数（艘） Number of Ships (unit)	项目数（个） Number of Items (unit)	实际获得数据（个） Quantity of Data Actually Obtained (unit)	发布通（公、简）报量（期） Quantity of Circulars (Bulletin, Brief Reports) (number)	提交报告数（份） Number of Reports Submitted (unit)
合　计 Total	**1 594**	**112**	**4 051**	**654 143**	**118**	**155**
大洋调查 Oceanic Survey	285	1	40		27	20
极地调查 Polar Survey	20	1	14	1 011		1
专项调查 Special Survey	399	16	26	29 229	5	33
其他调查 Other Survey	890	94	3 971	623 903	86	101

9-8 涉外海洋科学研究审批情况
Examination and Approval on Foreign-Related Marine Scientific Research

审批单位 Examing and Approving Units	审批研究活动申请（份） Application for the Examing and Approving Research Activity (unit)	船只作业计划审批（份） Examination and Approval of the Ship Operating Plan (unit)
国家海洋局 State Oceanic Administration , People's Republic of China	8	8

9-9 海洋档案及利用情况
Marine File and Its Use

指　标 Item	指标值 Data
室（馆）存档案 Files Deposited in the Archives	
纸介质（卷、册） Paper Media (reel,volume)	113 621
磁介质（盘） Magnetic Media (spool)	9 122
科技档案利用 Utilization of Scientific as Technological Files	
接待读者（人次） Readers Received (person-time)	2 596
借阅案卷、盘 Borrowed Files and Spools	
纸介质（卷次） Paper Media (Number of Reels)	3 239
电子（盘次） Electronic document (spool-times)	255

9-10 卫星遥感接收应用情况
Remote-Sensing Receiving and Utilization

指　标 Item	指标值 Data
卫星接收次数（次） Number of Satellite Receptions (time)	167 918
全年接收时间（分钟） Whole Year Receiving Time (minute)	5 371 825
全年实际接收存档数据量（GB） Amount of Data Actually Received and Placed on File in the whole year (GB)	28 734.81
累计存档数据量（GB） Total Amount of Data Placed on File (GB)	39 013.12
卫星数据分发 Satellite Data Distribution	
类别用户（个） Classified Users (number)	128
分发数据量（GB） Amount of Data Distributed (GB)	14 949.82

9-11 海洋标准化监督管理情况
Supervision and Management of Marine Standardization

单位：项 (item)

指　标 Item	指 标 值 Data
标准立项审查 Examination of the Standards for Authorization	
国家标准 National Standards	9
行业标准 Professional Standards	55
标准审查 Examination of Standards	
国家标准 National Standards	79
行业标准 Professional Standards	145
标准出版 Standards Publication	
国家标准 National Standards	5
行业标准 Professional Standards	5
标准实施监督检查（次） Implementation Supervision and Examination of Standards (time)	3

主要统计指标解释

1. 海域使用检查 针对不同类型的用海行为进行的监督检查。

2. 涉外海洋科研项目检查 主要针对国际组织、外国组织和个人为和平目的，单独或者与中华人民共和国的组织合作，使用船舶或者其他运载工具、设施，在中华人民共和国内海、领海以及中华人民共和国管辖的其他海域内进行的对海洋环境和海洋资源等的调查研究活动进行监督检查。

3. 海底电缆管道检查 主要是针对铺设海底电缆管道路由调查、铺设施工和维修改造等的监督检查。

4. 海洋工程建设项目环境保护检查 主要是针对防治海洋工程建设项目对海洋环境的污染损害的监督检查。

5. 海洋倾废检查 主要是针对防治倾倒废弃物对海洋环境的污染损害的监督检查。

6. 海洋生态保护检查 主要是针对红树林、珊瑚礁、滨海湿地、海岛、海湾、入海河口、重要渔业水域等具有典型性、代表性的海洋生态系统，珍稀、濒危海洋生物的天然集中分布区，具有重要经济价值的海洋生物生存区域及有重大科学文化价值的海洋自然历史遗迹和自然景观等海洋自然保护区以及其他需要予以特殊保护的区域的监督检查。

7. 断面监测 按照国家海洋局“断面监测方案”，每年定期利用船舶在沿海设定的断面上进行海洋水文、气象、生物、化学等项目的监测活动。

8. 浮标监测 在海上固定站位获取长期、连续海洋环境观测资料的海上锚定资料浮标。

9. 船舶监测 利用在固定航线上走航的商船或渔船，每日定时所在地的海洋环境状况（主要是水文气象要素），并将监测数据实时发送给有关单位，供海洋环境预报使用。

10. 大洋调查 以大洋科考、研究为目的的远洋调查。

11. 专项调查 为完成国家专项任务进行的海洋调查。

12. 全年接收时间 指全年整个应用系统接收的各类遥感卫星数据时，接收设备工作时间。

13. 全年实际接收存档数据量 指全年整个应用系统接收的各类遥感卫星数据原始数据量。

14. 累计存档数据量 指全年整个应用系统制作的各类遥感产品数据量。

15. 分发数据量 指应用系统提供数据产品用于开展各类应用的数据量。

16. 卫星接收次数 地面观测系统接收卫星数据的数量。

17. 国家标准 针对海洋领域内需要在全国范围内统一的有关技术要求所制定的国家标准。海洋国家标准由国家标准化主管部门统一批准、编号和发布。

18. 行业标准 对没有海洋国家标准而又需要在海洋领域内统一的技术要求所制定的标准。海洋行业标准由国家海洋局统一批准、编号和发布。

19. 获奖成果数 指本年度内机构作为第一完成单位从地（市）以上政府科技管理部门获得的各种科技成果奖的项数。

20. 国家级奖励 指国家自然科学奖、国家发明奖、国家科技进步奖。

21. 省部级奖励 指以国务院各部门名义颁发的或以省、自治区、直辖市政府（科委）名义颁发的重大科技成果奖和科技进步奖等。

22. 地市级奖励 指以地（市、州）政府（科委）和省属各部门名义颁发的重大科技成果奖和科技进步奖等。

Explanatory Notes on Main Statistical Indicators

1. Sea Area Use Supervision and inspection refers to the various types of sea area use conducts.

2. Inspection of Foreign-Related Marine Scientific Research Projects refers to the supervision and inspection of the activities of surveying marine environment and resources conducted by international organizations, foreign organizations and individuals for peaceful purposes, alone or in cooperation with PRC organizations, by using ships or other means of delivery as facilities in the internal seas and territorial waters of the People's Republic of China as well as in the other water under the jurisdiction of the People's Republic of China.

3. Inspection of Submarine Cables and Pipelines is mainly aimed at the survey of the submarine cables and pipelines laying route and supervision and inspection of the laying construction, repair and transformation.

4. Environmental Protection Inspection for the Marine Engineering Construction Projects is mainly directed against preventing and controlling the pollution damage of the marine engineering construction project to the marine environment.

5. Inspection of Oceanic Dumping of Wastes mainly refers to the supervision and inspection aimed at preventing and controlling the pollution damage of wastes dumping to the marine environment.

6. Inspection of Marine Ecological Protection mainly refers to the supervision and inspection of the typical and representative marine ecosystems such as mangroves, coral reef, coastal wetland, sea island, bay, estuaries open to the sea, important fishery waters, etc, the natural centralized distribution zones of rare and endangered marine life, the living areas for the marine life with significant economic values and the marine nature reserves such as the marine natural and historical remains and natural landscapes with important scientific and cultural values as well as other areas needing to be specially protected.

7. Sectional Monitoring According to the *"Sectional Monitoring Plan"* of the State Oceanic Administration, monitoring activities concerning such items as marine hydrology, meteorology, biology and chemistry are carried out regularly every year on the sections set in the coastal area.

8. Buoy Monitoring refers to the monitoring carried out by the offshore mooring data buoys which acquire long-term, continuous marine environmental observations at the fixed stations at sea.

9. Ship Monitoring Monitoring the marine environmental condition in the sea area of fixed times every day by using the merchant or fishing vessels cruising on the fixed navigation line (mainly the hydro meteorological elements) and transmitting the monitored data in real time to the related units for use in the marine environmental forecast.

10. Oceanic Survey refers to the oceanic surveys aimed at the oceanic scientific investigations and research.

11. Special-Subject Survey refers to the oceanic investigation for the purpose of fulfilling the state′s special tasks

12. Whole Year Receiving Time refers to the operating time of receiving equipment in the whole year when the whole application system receives all types of remote sensing satellite data.

13. Amount of Data Actually Received and Placed on File Throughout the year refers to the raw data of remote-sensing satellite data of various kinds received by the whole application system throughout the year.

14. Total Amount of Date Placed on File refers to the date amount of various remote-sensing products made by the whole application systems throughout the year.

15. Amount of Data Distributed refers to the amount of data products provided by the application system for various uses.

16. Number of Satellite Receptions refers to the amount of data received by the ground observation system.

17. National Standards refers to the standards formulated in view of the relevant technical requirements in the marine field that need to be unified throughout the country. The Marine National standards are approved, numbered and issued uniformly by the state department responsible for standardization.

18. Professional Standards refers to the standards formulated for the technical requirements which have no national standards but need to be unified in the marine field. The Marine Professional Standards are approved, numbered and issued by the State Oceanic Administration.

19. Number of Prize-Winning Results refers to the number of items of various scientific and technological results awards obtained from the departments of scientific and technological management of the governments above the prefecture (city) level by the institution and the first accomplishing unit.

20. State-Level Awards refers to the State Natural Science Award, State Invention Prize and State Scientific and Technological Progress Prize.

21. Province(Ministry)-Level Awards refers to the significant scientific and technological achievement prize and the scientific and technological progress prize awarded in the name of the departments of the State Council or the governments of provinces, autonomous regions and municipalities directly under the Central Government (commissions or science and technology).

22. Prefecture-Level Awards refers to the significant scientific and technological achievements prize and the scientific and technological progress prize awarded in the name of the prefecture government (commission of science and technology) and the various departments of the province.

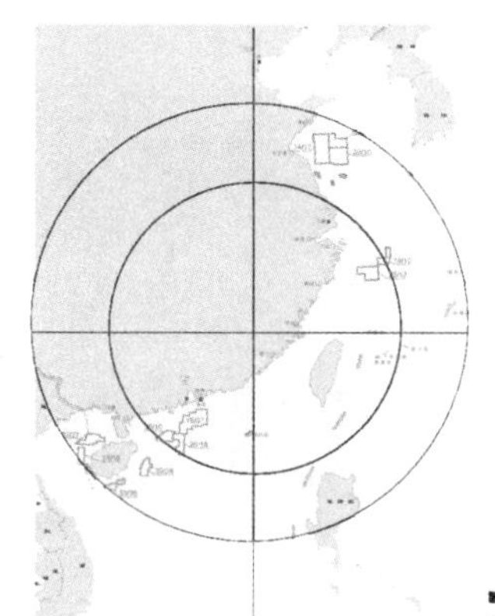

10

全国及沿海社会经济

National and Coastal Socioeconomy

10-1 国内生产总值
Gross Domestic Product

单位：亿元 (100 million yuan)

年 份 Year	国内生产总值 Gross Domestic Product			
		第一产业 Primary Industry	第二产业 Secondary Industry	第三产业 Tertiary Industry
2001	109 655.2	15 781.3	49 512.3	44 361.6
2002	120 332.7	16 537.0	53 896.8	49 898.9
2003	135 822.8	17 381.7	62 436.3	56 004.7
2004	159 878.3	21 412.7	73 904.3	64 561.3
2005	183 217.5	22 420.0	87 364.6	73 432.9
2006	216 314.4	24 040.0	103 719.5	88 554.9
2007	265 810.3	28 627.0	125 831.4	111 351.9
2008	314 045.4	33 702.0	149 003.4	131 340.0
2009	340 903.0	35 226.0	157 639.0	148 038.0

10-2 国内生产总值增长速度
Growth Rate of Gross Domestic Product

年 份 Year	国内生产总值(%) Gross Domestic Product(%)			
		第一产业 Primary Industry	第二产业 Secondary Industry	第三产业 Tertiary Industry
2001	8.3	2.8	8.4	10.3
2002	9.1	2.9	9.8	10.4
2003	10.0	2.5	12.7	9.5
2004	10.1	6.3	11.1	10.1
2005	11.3	5.2	12.1	12.2
2006	12.7	5.0	13.4	14.1
2007	14.2	3.7	15.1	16.0
2008	9.6	5.4	9.9	10.4
2009	9.1	4.2	9.9	9.3

注：本表按可比价格计算(上年为基期)。

Note:This table is calculated at the comparable price (with the previous year as the base period).

10-3 沿海地区生产总值
Gross Regional Product of Coastal Regions

单位：亿元 (100 million yuan)

地 区 Region	地区生产总值 Gross Regional Product	第一产业 Primary Industry	第二产业 Secondary Industry	第三产业 Tertiary Industry
合 计 Total	**207 493.03**	**15 630.18**	**105 482.45**	**86 380.40**
天 津 Tianjin	7 521.85	128.85	3 987.84	3 405.16
河 北 Hebei	17 235.48	2 207.34	8 959.83	6 068.31
辽 宁 Liaoning	15 212.49	1 414.90	7 906.34	5 891.25
上 海 Shanghai	15 046.45	113.82	6 001.78	8 930.85
江 苏 Jiangsu	34 457.30	2 261.86	18 566.37	13 629.07
浙 江 Zhejiang	22 990.35	1 163.08	11 908.49	9 918.78
福 建 Fujian	12 236.53	1 182.74	6 005.30	5 048.49
山 东 Shandong	33 896.65	3 226.64	18 901.83	11 768.18
广 东 Guangdong	39 482.56	2 010.27	19 419.70	18 052.59
广 西 Guangxi	7 759.16	1 458.49	3 381.54	2 919.13
海 南 Hainan	1 654.21	462.19	443.43	748.59

10-4 沿海地区生产总值增长速度
Growth Rate of Gross Regional Product of Coastal Regions

单位：%　　　　(%)

地　区 Region	2001	2002	2003	2004	2005	2006	2007	2008	2009
天　津 Tianjin	12.0	12.7	14.8	15.8	14.9	14.7	15.5	16.5	16.5
河　北 Hebei	8.7	9.6	11.6	12.9	13.4	13.4	12.8	10.1	10.0
辽　宁 Liaoning	9.0	10.2	11.5	12.8	12.7	14.2	15.0	13.4	13.1
上　海 Shanghai	10.5	11.3	12.3	14.2	11.4	12.7	15.2	9.7	8.2
江　苏 Jiangsu	10.2	11.7	13.6	14.8	14.5	14.9	14.9	12.7	12.4
浙　江 Zhejiang	10.6	12.6	14.7	14.5	12.8	13.9	14.7	10.1	8.9
福　建 Fujian	8.7	10.2	11.5	11.8	11.6	14.8	15.2	13.0	12.3
山　东 Shandong	10.0	11.7	13.4	15.4	15.0	14.7	14.2	12.0	12.2
广　东 Guangdong	10.5	12.4	14.8	14.8	14.1	14.8	14.9	10.4	9.7
广　西 Guangxi	8.3	10.6	10.2	11.8	13.1	13.6	15.1	12.8	13.9
海　南 Hainan	9.1	9.6	10.6	10.7	10.5	13.2	15.8	10.3	11.7

注：本表按可比价格计算(上年为基期)。

Note: This table is calculated at the comparable price (with the previous year as the base period).

10-5 沿海城市生产总值（2008年）
Gross Regional Product of Coastal Cities, 2008

单位：亿元 (100 million yuan)

沿海城市 Coastal City		地区生产总值 Gross Regional Product	第一产业 Primary Industry	第二产业 Secondary Industry	第三产业 Tertiary Industry
合 计	**Total**	**111 978.6**	**6 503.4**	**58 267.5**	**47 207.5**
天 津	**Tianjin**	**6 354.4**	**122.6**	**3 821.1**	**2 410.7**
河 北	**Hebei**	**6 086.3**	**632.5**	**3 308.2**	**2 145.6**
唐 山	Tangshan	3 561.2	340.0	2 113.3	1 107.9
秦皇岛	Qinhuangdao	809.0	91.1	328.0	389.9
沧 州	Cangzhou	1 716.1	201.4	866.9	647.8
辽 宁	**Liaoning**	**6 949.0**	**686.5**	**3 670.5**	**2 592.0**
大 连	Dalian	3 858.3	289.2	1 993.9	1 575.2
丹 东	Dandong	563.8	77.3	265.5	221.0
锦 州	Jinzhou	690.5	124.0	303.0	263.5
营 口	Yingkou	703.6	61.6	404.4	237.6
盘 锦	Panjin	675.0	68.9	489.1	117.0
葫芦岛	Huludao	457.8	65.5	214.6	177.7
上 海	**Shanghai**	**13 698.3**	**111.8**	**6 235.9**	**7 350.4**
江 苏	**Jiangsu**	**4 863.6**	**597.3**	**2 564.4**	**1 701.9**
南 通	Nantong	2 510.1	199.2	1 430.9	880.0
连云港	Lianyungang	750.2	122.8	355.1	272.3
盐 城	Yancheng	1 603.3	275.3	778.4	549.6
浙 江	**Zhejiang**	**17 663.4**	**827.6**	**9 551.2**	**7 284.6**
杭 州	Hangzhou	4 781.1	178.6	2 389.4	2 213.1
宁 波	Ningbo	3 964.1	167.4	2 196.7	1 600.0
温 州	Wenzhou	2 424.4	76.7	1 286.8	1 060.9
嘉 兴	Jiaxing	1 815.3	105.5	1 085.3	624.5
绍 兴	Shaoxing	2 223.0	116.7	1 329.1	777.2
舟 山	Zhoushan	490.2	49.2	226.4	214.6
台 州	Taizhou	1 965.3	133.5	1 037.5	794.3
福 建	**Fujian**	**8 704.2**	**773.1**	**4 517.6**	**3 413.5**
福 州	Fuzhou	2 284.1	234.9	1 083.9	965.3
厦 门	Xiamen	1 560.0	21.5	818.0	720.5
莆 田	Putian	610.0	75.1	344.9	190.0
泉 州	Quanzhou	2 705.3	120.3	1 605.6	979.4
漳 州	Zhangzhou	1 002.1	213.6	446.0	342.5
宁 德	Ningde	542.7	107.7	219.2	215.8

10-5 续表 continued

沿海城市 Coastal City		地区生产总值 Gross Regional Product	第一产业 Primary Industry	第二产业 Secondary Industry	第三产业 Tertiary Industry
山　东	**Shandong**	**16 205.4**	**1 188.5**	**9 634.5**	**5 382.4**
青　岛	Qingdao	4 436.2	223.4	2 255.5	1 957.3
东　营	Dongying	2 052.6	70.1	1 570.9	411.6
烟　台	Yantai	3 434.3	275.6	2 091.0	1 067.7
潍　坊	Weifang	2 491.9	281.7	1 455.1	755.1
威　海	Weihai	1 780.4	132.3	1 088.6	559.5
日　照	Rizhao	773.1	82.7	419.7	270.7
滨　州	Binzhou	1 236.9	122.7	753.7	360.5
广　东	**Guangdong**	**29 939.9**	**1 278.9**	**14 402.1**	**14 258.9**
广　州	Guangzhou	8 215.8	167.7	3 199.0	4 849.1
深　圳	Shenzhen	7 806.6	6.7	3 815.8	3 984.1
珠　海	Zhuhai	992.1	29.1	542.5	420.5
汕　头	Shantou	974.7	52.8	529.6	392.3
江　门	Jiangmen	1 280.6	103.4	737.5	439.7
湛　江	Zhanjiang	1 048.6	236.8	490.9	320.9
茂　名	Maoming	1 217.9	229.5	506.5	481.9
惠　州	Huizhou	1 290.5	90.7	759.5	440.3
汕　尾	Shanwei	350.2	62.9	162.3	125.0
阳　江	Yangjiang	483.9	115.5	196.2	172.2
东　莞	Dongguan	3 702.6	12.3	1 954.2	1 736.1
中　山	Zhongshan	1 408.5	44.3	850.6	513.6
潮　州	Chaozhou	442.8	33.9	251.4	157.5
揭　阳	Jieyang	725.1	93.3	406.1	225.7
广　西	**Guangxi**	**903.5**	**215.7**	**391.2**	**296.6**
北　海	Beihai	313.9	71.7	134.9	107.3
防城港	Fangchenggang	212.2	36.1	105.8	70.3
钦　州	Qinzhou	377.4	107.9	150.5	119.0
海　南	**Hainan**	**610.6**	**68.9**	**170.8**	**370.9**
海　口	Haikou	457.0	36.5	118.1	302.4
三　亚	Sanya	153.6	32.4	52.7	68.5

注：本表各省数据为合计数。

Note: The data for the provinces are the totals.

10-6 沿海县生产总值（2008年）
Gross Regional Product of Coastal Counties, 2008

单位：万元 (10 000 yuan)

沿海县 Coastal County		地区生产总值 Gross Regional Product	第一产业 Primary Industry	第二产业 Secondary Industry	第三产业 Tertiary Industry
合　计	**Total**	**239 953 740**	**32 539 633**	**129 652 214**	**77 761 893**
河　北	**Hebei**	**8 497 383**	**1 861 659**	**3 810 437**	**2 825 287**
滦　南	Luannan	2 252 899	458 956	1 111 315	682 628
乐　亭	Leting	2 025 646	488 379	822 190	715 077
唐　海	Tanghai	555 000	121 530	215 350	218 120
昌　黎	Changli	1 040 313	346 356	402 864	291 093
抚　宁	Funing	1 196 111	276 423	625 853	293 835
黄　骅	Huanghua	1 245 211	130 021	561 649	553 541
海　兴	Haixing	182 203	39 994	71 216	70 993
辽　宁	**Liaoning**	**19 031 076**	**3 947 572**	**9 592 010**	**5 491 494**
长　海	Changhai	345 500	256 052	48 215	41 233
瓦房店	Wafangdian	4 124 581	611 822	2 421 995	1 090 764
普兰店	Pulandian	3 189 435	537 422	1 755 105	896 908
庄　河	Zhuanghe	3 013 385	639 177	1 580 985	793 223
东　港	Donggang	2 570 207	369 538	1 300 601	900 068
凌　海	Linghai	1 301 978	313 224	636 510	352 244
盖　州	Gaizhou	983 918	204 918	426 000	353 000
大　洼	Dawa	1 313 514	377 796	665 519	270 199
盘　山	Panshan	906 252	282 252	439 000	185 000
绥　中	Suizhong	894 037	241 233	244 077	408 727
兴　城	Xingcheng	388 269	114 138	74 003	200 128
江　苏	**Jiangsu**	**29 221 814**	**4 439 281**	**15 203 842**	**9 578 691**
海　安	Hai'an	2 646 039	291 040	1 451 211	903 788
如　东	Rudong	2 634 042	340 940	1 413 059	880 043

10-6 续表1 continued

沿海县 Coastal County		地区生产总值 Gross Regional Product	第一产业 Primary Industry	第二产业 Secondary Industry	第三产业 Tertiary Industry
启 东	Qidong	3 270 011	416 212	1 749 151	1 104 648
通 州	Tongzhou	3 910 041	303 508	2 310 021	1 296 512
海 门	Haimen	3 760 981	278 681	2 274 300	1 208 000
赣 榆	Ganyu	1 407 100	271 700	667 900	467 500
东 海	Donghai	1 247 500	283 000	542 800	421 700
灌 云	Guanyun	967 200	294 900	423 700	248 600
灌 南	Guannan	792 500	167 500	401 300	223 700
响 水	Xiangshui	800 900	172 500	411 500	216 900
滨 海	Binhai	1 308 500	289 000	595 400	424 100
射 阳	Sheyang	1 732 300	420 100	732 500	579 700
东 台	Dongtai	2 671 100	496 900	1 273 700	900 500
大 丰	Dafeng	2 073 600	413 300	957 300	703 000
浙 江	**Zhejiang**	**58 800 184**	**3 888 770**	**34 176 155**	**20 735 259**
象 山	Xiangshan	2 206 172	333 010	1 097 339	775 823
宁 海	Ninghai	2 179 280	228 635	1 241 036	709 609
余 姚	Yuyao	4 847 113	271 752	2 928 941	1 646 420
慈 溪	Cixi	6 014 360	281 446	3 737 307	1 995 607
奉 化	Fenghua	1 879 386	166 336	954 560	758 490
洞 头	Dongtou	304 136	33 430	102 872	167 834
平 阳	Pingyang	1 620 327	87 944	796 013	736 370
苍 南	Cangnan	2 055 564	162 487	929 209	963 868
瑞 安	Rui'an	3 681 587	120 633	1 952 175	1 608 779
乐 清	Yueqing	4 042 164	144 530	2 518 744	1 378 890
海 盐	Haiyan	2 021 828	147 334	1 335 817	538 677
海 宁	Haining	3 489 469	171 520	2 179 153	1 138 796
平 湖	Pinghu	2 763 102	140 366	1 822 306	800 430
绍 兴	Shaoxing	6 082 566	216 369	3 886 210	1 979 987
上 虞	Shangyu	3 483 299	243 456	2 123 624	1 116 219
岱 山	Daishan	841 900	143 600	395 500	302 800

10-6 续表2 continued

沿海县 Coastal County		地区生产总值 Gross Regional Product	第一产业 Primary Industry	第二产业 Secondary Industry	第三产业 Tertiary Industry
嵊泗	Shengsi	558 600	86 200	280 900	191 500
玉环	Yuhuan	2 528 123	173 239	1 638 959	715 925
三门	Sanmen	818 864	138 209	353 207	327 448
温岭	Wenling	4 785 480	379 564	2 525 616	1 880 300
临海	Linhai	2 596 864	218 710	1 376 667	1 001 487
福建	**Fujian**	**35 033 400**	**4 251 500**	**19 173 000**	**11 608 900**
连江	Lianjiang	1 359 200	510 500	414 900	433 800
罗源	Luoyuan	706 500	147 500	422 600	136 400
平潭	Pingtan	604 900	202 800	97 900	304 200
福清	Fuqing	4 015 200	505 600	2 281 600	1 228 000
长乐	Changle	2 389 600	217 900	1 617 000	554 700
仙游	Xianyou	1 104 700	182 200	499 500	423 000
惠安	Hui'an	3 152 100	222 700	1 844 300	1 085 100
石狮	Shishi	2 878 000	132 600	1 544 600	1 200 800
晋江	Jinjiang	6 977 100	133 800	4 494 700	2 348 600
南安	Nan'an	3 641 300	173 800	2 265 200	1 202 300
云霄	Yunxiao	494 600	177 200	116 900	200 500
漳浦	Zhangpu	1 116 600	362 700	281 200	472 700
诏安	Zhao'an	687 500	249 200	225 900	212 400
东山	Dongshan	510 900	169 200	193 600	148 100
龙海	Longhai	2 444 300	335 800	1 521 700	586 800
霞浦	Xiapu	739 500	192 800	210 300	336 400
福安	Fu'an	1 306 300	188 700	717 200	400 400
福鼎	Fuding	905 100	146 500	423 900	334 700
山东	**Shandong**	**62 840 725**	**5 873 730**	**37 710 192**	**19 256 803**
胶州	Jiaozhou	4 740 546	318 763	2 725 600	1 696 183
即墨	Jimo	4 989 639	387 480	2 751 000	1 851 159

10-6 续表3 continued

沿海县 Coastal County		地区生产总值 Gross Regional Product	第一产业 Primary Industry	第二产业 Secondary Industry	第三产业 Tertiary Industry
胶　南	Jiaonan	4 396 319	354 032	2 583 100	1 459 187
垦　利	Kenli	1 601 087	111 208	1 169 329	320 550
利　津	Lijin	1 000 476	164 458	589 755	246 263
广　饶	Guangrao	3 223 317	248 011	2 380 030	595 276
长　岛	Changdao	349 575	196 209	38 668	114 698
龙　口	Longkou	5 700 409	279 800	3 610 780	1 809 829
莱　阳	Laiyang	2 803 085	300 656	1 655 777	846 652
莱　州	Laizhou	4 102 649	412 273	2 443 681	1 246 695
蓬　莱	Penglai	2 905 842	200 000	1 826 212	879 630
招　远	Zhaoyuan	3 694 690	201 184	2 402 922	1 090 584
海　阳	Haiyang	1 820 255	376 296	865 778	578 181
寿　光	Shouguang	4 006 320	515 498	2 168 362	1 322 460
昌　邑	Changyi	1 925 010	242 026	1 221 379	461 605
文　登	Wendeng	4 768 437	360 876	2 885 731	1 521 830
荣　成	Rongcheng	5 500 166	526 921	3 249 248	1 723 997
乳　山	Rushan	2 884 375	242 236	1 734 138	908 001
无　棣	Wudi	1 456 610	233 003	932 718	290 889
沾　化	Zhanhua	971 918	202 800	475 984	293 134
广　东	**Guangdong**	**18 601 370**	**4 792 183**	**7 764 968**	**6 044 219**
南　澳	Nan'ao	71 152	24 246	26 388	20 518
台　山	Taishan	1 826 341	260 553	1 036 033	529 755
恩　平	Enping	793 699	112 704	292 513	388 482
遂　溪	Suixi	989 256	492 624	233 499	263 133
徐　闻	Xuwen	629 665	323 680	72 667	233 318
廉　江	Lianjiang	1 341 023	542 829	441 540	356 654
雷　州	Leizhou	904 315	459 064	146 678	298 573
吴　川	Wuchuan	759 136	167 421	309 838	281 877
电　白	Dianbai	1 699 086	486 000	594 087	618 999

10-6 续表4 continued

沿海县 Coastal County		地区生产总值 Gross Regional Product	第一产业 Primary Industry	第二产业 Secondary Industry	第三产业 Tertiary Industry
惠 东	Huidong	2 010 896	249 058	1 126 071	635 767
海 丰	Haifeng	1 288 103	193 184	546 640	548 279
陆 丰	Lufeng	1 030 390	262 204	441 649	326 537
阳 西	Yangxi	682 610	279 974	200 511	202 125
阳 东	Yangdong	970 018	227 944	477 806	264 268
饶 平	Raoping	1 025 990	198 833	418 018	409 139
揭 东	Jiedong	1 658 902	240 310	974 690	443 902
惠 来	Huilai	920 788	271 555	426 340	222 893
广 西	**Guangxi**	**1 474 627**	**431 025**	**537 675**	**505 927**
合 浦	Hepu	1 149 905	364 229	459 651	326 025
东 兴	Dongxing	324 722	66 796	78 024	179 902
海 南	**Hainan**	**6 453 161**	**3 053 913**	**1 683 935**	**1 715 313**
琼 海	Qionghai	757 730	398 781	97 930	261 019
儋 州	Danzhou	1 035 771	578 156	168 236	289 379
文 昌	Wenchang	837 697	400 127	170 651	266 919
万 宁	Wanning	684 175	264 536	156 195	263 444
东 方	Dongfang	692 544	178 821	387 798	125 925
澄 迈	Chengmai	714 449	236 492	345 424	132 533
临 高	Lingao	567 086	426 275	36 599	104 212
昌 江	Changjiang	437 048	120 212	239 291	77 545
乐 东	Ledong	404 248	270 946	29 168	104 134
陵 水	Lingshui	322 413	179 567	52 643	90 203

注：本表各省数据为合计数，缺少上海崇明县，福建金门县数据。

Note: The data for the provinces are the totals, and lacking the data for Chongming of Shanghai Province; Jinmen of Fujian Province.

10-7 沿海地区财政收支
Financial Revenue and Expenditure by Coastal Regions

单位：万元 (10 000 yuan)

地 区 Region	一般预算收入 General Budgetary Revenue	一般预算支出 General Budgetary Expenditure
合 计 Total	**18 972.02**	**26 936.36**
天 津 Tianjin	821.99	1 124.28
河 北 Hebei	1 067.12	2 347.59
辽 宁 Liaoning	1 591.22	2 682.39
上 海 Shanghai	2 540.30	2 989.65
江 苏 Jiangsu	3 228.78	4 017.36
浙 江 Zhejiang	2 142.51	2 653.35
福 建 Fujian	932.43	1 411.82
山 东 Shandong	2 198.63	3 267.67
广 东 Guangdong	3 649.81	4 334.37
广 西 Guangxi	620.99	1 621.82
海 南 Hainan	178.24	486.06

10-8 沿海地区教育基本情况
Basic Conditions of Education by Coastal Regions

地　区 Region	高等学校数 （所） Institutions of Higher Education (unit)	本、专科在校学生数 （人） Number of Students Enrolled in Undergraduate or Specialized Courses (person)	本、专科毕（结）业生数 （人） Number of Students Graduated in Undergraduate or Specialized Courses (person)
全国总计 National Total	**2 035**	**21 446 570**	**5 311 023**
天　津 Tianjin	55	405 968	101 369
河　北 Hebei	109	1 060 450	282 705
辽　宁 Liaoning	107	852 467	206 211
上　海 Shanghai	66	512 809	126 925
江　苏 Jiangsu	148	1 653 427	412 672
浙　江 Zhejiang	99	866 496	218 226
福　建 Fujian	84	606 284	142 814
山　东 Shandong	126	1 592 974	431 598
广　东 Guangdong	125	1 334 089	309 190
广　西 Guangxi	68	528 342	121 457
海　南 Hainan	17	142 082	30 844

10-9 沿海地区卫生基本情况
Basic Conditions of Public Health by Coastal Regions

地　区 Region	卫生机构数 （个） Health Institutions (unit)	医疗机构床位数 （张） Total Beds (bed)	卫生机构人员 （人） Number of Employed Personnel in Health Institutions (person)
全国总计 National Total	**916 571**	**4 416 612**	**7 781 448**
天　津 Tianjin	4 238	46 353	93 366
河　北 Hebei	80 963	232 638	407 351
辽　宁 Liaoning	34 729	191 492	310 007
上　海 Shanghai	4 460	99 704	169 506
江　苏 Jiangsu	30 571	250 809	437 008
浙　江 Zhejiang	29 549	170 199	327 014
福　建 Fujian	26 613	104 290	184 866
山　东 Shandong	63 885	347 052	602 143
广　东 Guangdong	44 314	271 982	555 799
广　西 Guangxi	32 355	131 569	245 611
海　南 Hainan	4 661	23 526	49 898

10-10 沿海地区能源消耗指标
Indicators of Energy Consumption by Coastal Regions

地　区 Region	单位地区生产总值能耗（等价值）Energy Consumption Per Unit of GRP (Equivalent Value)		单位工业增加值能耗（规模以上，当量值）Energy Consumption Per Unit of Industrial Value-Added (above Designated Size, Equivalent Weight)	
	指标值（吨标准煤/万元）Index (ton of SCE/10 000 yuan)	上升或下降（±%）Change (±%)	指标值（吨标准煤/万元）Index (ton of SCE/10 000 yuan)	上升或下降（±%）Change (±%)
天　津 Tianjin	0. 836	-6. 03	0. 911	-13. 54
河　北 Hebei	1. 640	-5. 02	2. 999	-9. 54
辽　宁 Liaoning	1. 439	-5. 08	2. 257	-6. 95
上　海 Shanghai	0. 727	-6. 17	0. 957	-5. 00
江　苏 Jiangsu	0. 761	-5. 17	1. 107	-10. 17
浙　江 Zhejiang	0. 741	-5. 41	1. 123	-4. 96
福　建 Fujian	0. 811	-3. 81	1. 150	-2. 70
山　东 Shandong	1. 072	-5. 46	1. 543	-9. 20
广　东 Guangdong	0. 684	-4. 27	0. 809	-6. 94
广　西 Guangxi	1. 057	-4. 43	2. 235	-6. 68
海　南 Hainan	0. 850	-2. 81	2. 613	-4. 53

10-10 续表 continued

地区 Region	单位地区生产总值电耗（等价值） Electricity Consumption per Unit of GRP (equivalent value)	
	指标值（千瓦小时/万元） Index (kW·h/10 000 yuan)	上升或下降（±%） Change (±%)
天津 Tianjin	782.88	-8.49
河北 Hebei	1 449.94	-2.52
辽宁 Liaoning	1 119.99	-6.82
上海 Shanghai	808.49	-6.39
江苏 Jiangsu	1 064.25	-5.50
浙江 Zhejiang	1 176.50	-2.33
福建 Fujian	1 032.05	-5.87
山东 Shandong	972.49	-3.86
广东 Guangdong	1 002.09	-6.13
广西 Guangxi	1 279.87	-2.00
海南 Hainan	922.89	-2.61

注：地区生产总值和工业增加值按2005年价格计算。

Note: Gross regional product and industrial value-added are caculated at 2005 constant prices.

10-11 沿海地区用水情况
Water Use of Coastal Regions

地 区 Region	全年供水总量 （亿立方米） Total Annual Volume of Water Supply (100 million m^3)	人均日生活用水量 （升） Per Capita Daily Consumption of Tap Water for Residential Use (liter)
全国总计 National Total	**496.746 7**	**176.6**
天 津 Tianjin	7.013 8	133.2
河 北 Hebei	15.699 1	124.8
辽 宁 Liaoning	28.873 2	124.2
上 海 Shanghai	34.138 9	207.0
江 苏 Jiangsu	44.903 7	207.2
浙 江 Zhejiang	26.980 9	201.6
福 建 Fujian	13.426 4	191.9
山 东 Shandong	27.559 2	129.9
广 东 Guangdong	78.507 1	253.1
广 西 Guangxi	13.898 1	259.1
海 南 Hainan	3.117 6	265.1

10-12 沿海地区全社会固定资产投资
Total Investment in Fixed Assets by the Whole Society of Coastal Regions

单位：亿元　(100 million yuan)

地　区 Region	全社会固定资产投资 Total Investment in Fixed Assets in the Whole Country	#农、林、牧、渔业 Agriculture,Forestry, Animal Husbandry and Fishery	#交通运输、仓储和邮政业 Transport,Storage and Post
全国总计 National Total	**224 598.8**	**6 894.9**	**24 974.7**
天　津 Tianjin	4 738.2	77.1	483.7
河　北 Hebei	12 269.8	509.9	1 026.2
辽　宁 Liaoning	12 292.5	322.6	757.6
上　海 Shanghai	5 043.8	11.4	882.8
江　苏 Jiangsu	18 949.9	177.1	1 020.2
浙　江 Zhejiang	10 742.3	98.2	1 008.7
福　建 Fujian	6 231.2	124.1	885.4
山　东 Shandong	19 034.5	512.7	1 032.5
广　东 Guangdong	12 933.1	149.2	1 596.2
广　西 Guangxi	5 237.2	218.5	602.3
海　南 Hainan	988.3	31.0	186.4

10-13 沿海地区按经营单位所在地分货物进出口总额
Total Value of Imports and Exports by Location of Importers/Exporters of Coastal Regions

单位：万美元 (10 000 US$)

地　区 Region	进出口 Total	出口 Exports	进口 Imports
全国总计 National Total	**220 753 500**	**120 161 181**	**100 592 320**
天　津 Tianjin	6 383 123	2 989 272	3 393 852
河　北 Hebei	2 962 725	1 568 890	1 393 835
辽　宁 Liaoning	6 293 438	3 341 493	2 951 945
上　海 Shanghai	27 771 361	14 179 603	13 591 758
江　苏 Jiangsu	33 873 970	19 919 919	13 954 051
浙　江 Zhejiang	18 773 086	13 301 295	5 471 791
福　建 Fujian	7 964 959	5 331 911	2 633 048
山　东 Shandong	13 905 337	7 949 071	5 956 266
广　东 Guangdong	61 109 405	35 895 489	25 213 916
广　西 Guangxi	1 425 473	837 537	587 936
海　南 Hainan	488 163	130 863	357 300

10-14 沿海地区城镇居民平均每人全年家庭收入和消费性支出
Per Capita Annual Income and Consumption Expenditure of Urban Households by Coastal Regions

单位：元 (yuan)

地　区 Region	总收入 Total Income	可支配收入 Disposable Income	消费性支出 Consumption Expenditure
全国平均水平 National Average	**18 858.09**	**17 174.65**	**12 264.55**
天　津 Tianjin	23 565.67	21 402.01	14 801.35
河　北 Hebei	15 675.75	14 718.25	9 678.75
辽　宁 Liaoning	17 757.70	15 761.38	12 324.58
上　海 Shanghai	32 402.97	28 837.78	20 992.35
江　苏 Jiangsu	22 494.94	20 551.72	13 153.00
浙　江 Zhejiang	27 119.30	24 610.81	16 683.48
福　建 Fujian	21 692.35	19 576.83	13 450.57
山　东 Shandong	19 336.91	17 811.04	12 012.73
广　东 Guangdong	24 116.46	21 574.72	16 857.50
广　西 Guangxi	17 032.89	15 451.48	10 352.38
海　南 Hainan	14 909.28	13 750.85	10 086.65

10-15 沿海地区农村居民家庭人均纯收入和生活消费支出
Per Capita Annual Net Income and Consumption Expenditure of Rural Households by Coastal Regions

单位：元 (yuan)

地　区 Region	纯收入 Net Income	生活消费支出 Consumption Expenditure
全国平均水平 National Average	**5 153.17**	**3 993.45**
天　津 Tianjin	8 687.56	4 273.15
河　北 Hebei	5 149.67	3 349.74
辽　宁 Liaoning	5 958.00	4 254.03
上　海 Shanghai	12 482.94	9 804.37
江　苏 Jiangsu	8 003.54	5 804.45
浙　江 Zhejiang	10 007.31	7 731.70
福　建 Fujian	6 680.18	5 015.72
山　东 Shandong	6 118.77	4 417.18
广　东 Guangdong	6 906.93	5 019.81
广　西 Guangxi	3 980.44	3 231.14
海　南 Hainan	4 744.36	3 088.56

10-16 沿海地区人口情况
Population of Coastal Regions

单位：万人　　(10 000 persons)

地　区 Region	年末总人口 Total Population by the End of the Year	城镇人口 Urban Population
全国总计 National Total	**133 474**	**62 186**
天　津 Tianjin	1 228	958
河　北 Hebei	7 034	3 025
辽　宁 Liaoning	4 319	2 607
上　海 Shanghai	1 921	1 702
江　苏 Jiangsu	7 725	4 295
浙　江 Zhejiang	5 180	2 999
福　建 Fujian	3 627	1 864
山　东 Shandong	9 470	4 576
广　东 Guangdong	9 638	6 110
广　西 Guangxi	4 856	1 904
海　南 Hainan	864	424

注：本表数据根据2009年人口变动情况抽样调查数据推算。
Note: Data in the table are estimates from the 2009 National Sample Survey on Population Changes.

10-17 沿海城市人口情况（2008年）
Population of Coastal Cities, 2008

单位：万人 (10 000 persons)

沿海城市 Coastal City		年末总人口 Total Population by the End of the Year
合　计	**Total**	**24 035.7**
天　津	**Tianjin**	**1 176.0**
河　北	**Hebei**	**1 725.4**
唐　山	Tangshan	729.4
秦皇岛	Qinhuangdao	285.9
沧　州	Cangzhou	710.1
辽　宁	**Liaoning**	**1 779.7**
大　连	Dalian	583.4
丹　东	Dandong	242.7
锦　州	Jinzhou	310.2
营　口	Yingkou	233.8
盘　锦	Panjin	129.2
葫芦岛	Huludao	280.4
上　海	**Shanghai**	**1 391.0**
江　苏	**Jiangsu**	**2 063.7**
南　通	Nantong	763.7
连云港	Lianyungang	488.3
盐　城	Yancheng	811.7

10-17 续表1 continued

沿海城市 Coastal City		年末总人口 Total Population by the End of the Year
浙　江	**Zhejiang**	**3 463.8**
杭　州	Hangzhou	677.6
宁　波	Ningbo	568.1
温　州	Wenzhou	772.0
嘉　兴	Jiaxing	338.1
绍　兴	Shaoxing	437.1
舟　山	Zhoushan	96.8
台　州	Taizhou	574.1
福　建	**Fujian**	**2 607.5**
福　州	Fuzhou	636.0
厦　门	Xiamen	173.7
莆　田	Putian	316.6
泉　州	Quanzhou	677.7
漳　州	Zhangzhou	468.5
宁　德	Ningde	335.0
山　东	**Shandong**	**3 372.2**
青　岛	Qingdao	761.6
东　营	Dongying	184.0
烟　台	Yantai	651.7
潍　坊	Weifang	862.5
威　海	Weihai	252.2
日　照	Rizhao	284.5
滨　州	Binzhou	375.7

10-17 续表2 continued

沿海城市	Coastal City	年末总人口 Total Population by the End of the Year
广　东	**Guangdong**	**5 639.0**
广　州	Guangzhou	784.2
深　圳	Shenzhen	232.5
珠　海	Zhuhai	99.5
汕　头	Shantou	506.6
江　门	Jiangmen	389.9
湛　江	Zhanjiang	753.9
茂　名	Maoming	725.7
惠　州	Huizhou	318.8
汕　尾	Shanwei	336.0
阳　江	Yangjiang	273.3
东　莞	Dongguan	174.9
中　山	Zhongshan	146.4
潮　州	Chaozhou	256.1
揭　阳	Jieyang	641.2
广　西	**Guangxi**	**607.0**
北　海	Beihai	157.7
防城港	Fangchenggang	84.8
钦　州	Qinzhou	364.5
海　南	**Hainan**	**210.4**
海　口	Haikou	155.8
三　亚	Sanya	54.6

注：本表各省数据为合计数。

Note: The data for the provinces are the totals.

10-18 沿海县人口情况（2008年）
Population of Coastal Counties, 2008

单位：万人 (10 000 persons)

沿海县	Coastal County	年末总人口 Total Population by the End of the Year
合　计	**Total**	**8 610.26**
河　北	**Hebei**	**295.68**
滦　南	Luannan	58.20
乐　亭	Leting	49.71
唐　海	Tanghai	14.16
昌　黎	Changli	55.43
抚　宁	Funing	52.35
黄　骅	Huanghua	43.26
海　兴	Haixing	22.57
辽　宁	**Liaoning**	**661.40**
长　海	Changhai	7.40
瓦房店	Wafangdian	102.50
普兰店	Pulandian	82.80
庄　河	Zhuanghe	92.30
东　港	Donggang	61.20
凌　海	Linghai	53.50
盖　州	Gaizhou	73.20
大　洼	Dawa	39.50
盘　山	Panshan	29.50
绥　中	Suizhong	64.10
兴　城	Xingcheng	55.40
上　海	**Shanghai**	
崇　明	Chongming	
江　苏	**Jiangsu**	**1 402.82**

10-18 续表1 continued

沿海县 Coastal County		年末总人口 Total Population by the End of the Year
海　安	Hai'an	93.81
如　东	Rudong	105.67
启　东	Qidong	111.41
通　州	Tongzhou	124.27
海　门	Haimen	100.12
赣　榆	Ganyu	109.95
东　海	Donghai	111.79
灌　云	Guanyun	110.35
灌　南	Guannan	75.28
响　水	Xiangshui	60.76
滨　海	Binhai	115.72
射　阳	Sheyang	96.64
东　台	Dongtai	114.60
大　丰	Dafeng	72.45
浙　江	**Zhejiang**	**1 453.23**
象　山	Xiangshan	53.51
宁　海	Ninghai	60.07
余　姚	Yuyao	83.11
慈　溪	Cixi	103.12
奉　化	Fenghua	48.16
洞　头	Dongtou	12.63
平　阳	Pingyang	85.65
苍　南	Cangnan	126.53
瑞　安	Rui'an	117.52
乐　清	Yueqing	120.91
海　盐	Haiyan	36.92
海　宁	Haining	65.09
平　湖	Pinghu	48.44
绍　兴	Shaoxing	71.46
上　虞	Shangyu	77.31

10-18 续表2 continued

沿海县	Coastal County	年末总人口 Total Population by the End of the Year
岱　山	Daishan	19.20
嵊　泗	Shengsi	7.99
玉　环	Yuhuan	41.10
三　门	Sanmen	42.27
温　岭	Wenling	117.58
临　海	Linhai	114.66
福　建	**Fujian**	**1 264.49**
连　江	Lianjiang	62.37
罗　源	Luoyuan	25.37
平　潭	Pingtan	38.48
福　清	Fuqing	124.83
长　乐	Changle	67.03
仙　游	Xianyou	105.72
惠　安	Hui'an	94.60
石　狮	Shishi	31.35
晋　江	Jinjiang	105.04
南　安	Nan'an	149.67
云　霄	Yunxiao	42.76
漳　浦	Zhangpu	83.65
诏　安	Zhao'an	58.84
东　山	Dongshan	20.63
龙　海	Longhai	80.39
霞　浦	Xiapu	52.06
福　安	Fu'an	64.36
福　鼎	Fuding	57.34
山　东	**Shandong**	**1 217.20**
胶　州	Jiaozhou	79.60
即　墨	Jimo	112.10

10-18 续表3 continued

沿海县 Coastal County		年末总人口 Total Population by the End of the Year
胶　南	Jiaonan	83.40
垦　利	Kenli	21.80
利　津	Lijin	29.70
广　饶	Guangrao	49.30
长　岛	Changdao	4.30
龙　口	Longkou	63.00
莱　阳	Laiyang	87.50
莱　州	Laizhou	85.90
蓬　莱	Penglai	44.80
招　远	Zhaoyuan	56.90
海　阳	Haiyang	66.70
寿　光	Shouguang	102.50
昌　邑	Changyi	57.90
文　登	Wendeng	64.20
荣　成	Rongcheng	66.70
乳　山	Rushan	57.40
无　棣	Wudi	44.70
沾　化	Zhanhua	38.80
广　东	**Guangdong**	**1 681.75**
南　澳	Nan'ao	7.27
台　山	Taishan	98.44
恩　平	Enping	50.07
遂　溪	Suixi	102.60
徐　闻	Xuwen	71.43
廉　江	Lianjiang	161.85
雷　州	Leizhou	161.54
吴　川	Wuchuan	106.94
电　白	Dianbai	138.29

10-18 续表4 continued

沿海县 Coastal County		年末总人口 Total Population by the End of the Year
惠　东	Huidong	79.41
海　丰	Haifeng	82.13
陆　丰	Lufeng	171.20
阳　西	Yangxi	50.04
阳　东	Yangdong	47.07
饶　平	Raoping	100.02
揭　东	Jiedong	125.93
惠　来	Huilai	127.52
广　西	**Guangxi**	**110.40**
东　兴	Dongxing	11.98
合　浦	Hepu	98.42
海　南	**Hainan**	**523.29**
琼　海	Qionghai	48.25
儋　州	Danzhou	103.07
文　昌	Wenchang	57.08
万　宁	Wanning	59.10
东　方	Dongfang	42.72
澄　迈	Chengmai	52.88
临　高	Lingao	47.34
昌　江	Changjiang	25.47
乐　东	Ledong	51.61
陵　水	Lingshui	35.77

注：本表各省数据为合计数，缺少福建金门县数据。

Note: The data for the provinces are the totals, and lacking the data for Jinmen of Fujian Province.

10-19 沿海地区就业人员情况
Number of Employed Persons by Coastal Regions

单位：万人 (10 000 persons)

地 区 Region	就业人员 Number of Employed Persons	城镇就业人员 Urban Employed Persons
全国总计 National Total	**77 995.0**	**31 120.0**
天 津 Tianjin	507.3	315.6
河 北 Hebei	3 899.7	955.4
辽 宁 Liaoning	2 190.0	1 009.4
上 海 Shanghai	929.2	723.5
江 苏 Jiangsu	4 536.1	1 868.4
浙 江 Zhejiang	3 825.2	1 503.8
福 建 Fujian	2 168.9	793.5
山 东 Shandong	5 449.8	1 459.1
广 东 Guangdong	5 643.3	2 277.2
广 西 Guangxi	2 862.6	521.6
海 南 Hainan	431.4	149.9

10-20 沿海城市就业人员情况（2008年）
Number of Employed Persons by Coastal Cities, 2008

单位：万人 (10 000 persons)

沿海城市 Coastal City		就业人员 Number of Employed Persons	城镇就业人员 Urban Employed Persons
合　计	**Total**	**15 577.80**	**6 184.50**
天　津	**Tianjin**	**647.30**	**477.70**
河　北	**Hebei**	**956.40**	**266.10**
唐　山	Tangshan	412.00	132.20
秦皇岛	Qinhuangdao	160.60	52.80
沧　州	Cangzhou	383.80	81.10
辽　宁	**Liaoning**	**1 009.50**	**489.30**
大　连	Dalian	371.00	228.90
丹　东	Dandong	118.10	40.80
锦　州	Jinzhou	156.60	56.70
营　口	Yingkou	126.00	54.20
盘　锦	Panjin	104.70	68.10
葫芦岛	Huludao	133.10	40.60
上　海	**Shanghai**	**1 053.20**	**377.20**
江　苏	**Jiangsu**	**1 067.10**	**287.10**
南　通	Nantong	454.90	111.40
连云港	Lianyungang	276.10	61.10
盐　城	Yancheng	336.10	114.60
浙　江	**Zhejiang**	**2 594.50**	**1 168.90**
杭　州	Hangzhou	569.20	304.10
宁　波	Ningbo	439.90	230.00
温　州	Wenzhou	536.80	261.70
嘉　兴	Jiaxing	294.40	116.00
绍　兴	Shaoxing	315.40	130.70
舟　山	Zhoushan	63.20	22.90
台　州	Taizhou	375.60	103.50
福　建	**Fujian**	**1 611.60**	**600.10**
福　州	Fuzhou	365.90	153.30
厦　门	Xiamen	165.60	144.80
莆　田	Putian	166.50	36.80

10-20 续表 continued

沿海城市 Coastal City		就业人员 Number of Employed Persons	城镇就业人员 Urban Employed Persons
泉　州	Quanzhou	494.50	184.10
漳　州	Zhangzhou	258.80	52.30
宁　德	Ningde	160.30	28.80
山　东	**Shandong**	**2 008.80**	**667.00**
青　岛	Qingdao	488.20	217.50
东　营	Dongying	106.50	47.80
烟　台	Yantai	397.40	141.60
潍　坊	Weifang	479.90	120.70
威　海	Weihai	151.60	63.00
日　照	Rizhao	164.40	31.50
滨　州	Binzhou	220.80	44.90
广　东	**Guangdong**	**4 157.10**	**1 728.80**
广　州	Guangzhou	714.50	366.20
深　圳	Shenzhen	670.40	431.90
珠　海	Zhuhai	101.50	84.40
汕　头	Shantou	238.80	83.80
江　门	Jiangmen	233.80	79.30
湛　江	Zhanjiang	312.60	66.40
茂　名	Maoming	332.90	54.60
惠　州	Huizhou	241.10	158.70
汕　尾	Shanwei	119.90	50.70
阳　江	Yangjiang	156.20	41.60
东　莞	Dongguan	439.20	89.50
中　山	Zhongshan	203.80	104.40
潮　州	Chaozhou	136.90	21.70
揭　阳	Jieyang	255.50	95.60
广　西	**Guangxi**	**349.40**	**48.10**
北　海	Beihai	75.60	10.30
防城港	Fangchenggang	50.70	13.30
钦　州	Qinzhou	223.10	24.50
海　南	**Hainan**	**122.90**	**74.20**
海　口	Haikou	96.20	62.50
三　亚	Sanya	26.70	11.70

注：本表各省数据为合计数。

Note: The data for the provinces are the totals.

10-21 沿海县就业人员情况（2008年）
Number of Employed Persons of Coastal Counties, 2008

单位：人 (person)

沿海县 Coastal County		城镇单位在岗职工人数 Urban Employed Persons	乡村从业人员 Rural Laborer
合　计	**Total**	**6 636 118**	**38 619 243**
河　北	**Hebei**	**180 285**	**1 414 859**
滦　南	Luannan	27 785	288 028
乐　亭	Leting	21 012	259 240
唐　海	Tanghai	40 743	69 221
昌　黎	Changli	20 532	281 228
抚　宁	Funing	29 515	243 511
黄　骅	Huanghua	29 126	169 884
海　兴	Haixing	11 572	103 747
辽　宁	**Liaoning**	**580 324**	**2 692 703**
长　海	Changhai	8 504	29 127
瓦房店	Wafangdian	143 885	374 069
普兰店	Pulandian	41 834	309 118
庄　河	Zhuanghe	38 392	385 444
东　港	Donggang	24 600	262 060
凌　海	Linghai	20 247	227 109
盖　州	Gaizhou	15 726	316 879
大　洼	Dawa	183 765	167 271
盘　山	Panshan	70 859	163 984
绥　中	Suizhong	15 957	269 019
兴　城	Xingcheng	16 555	188 623
江　苏	**Jiangsu**	**701 827**	**5 891 100**
海　安	Hai'an	67 582	405 300

10-21 续表1 continued

沿海县 Coastal County	城镇单位在岗职工人数 Urban Employed Persons	乡村从业人员 Rural Laborer
如　东　Rudong	62 339	498 700
启　东　Qidong	58 534	586 700
通　州　Tongzhou	69 060	592 400
海　门　Haimen	62 327	503 100
赣　榆　Ganyu	34 624	410 100
东　海　Donghai	36 981	454 800
灌　云　Guanyun	43 548	396 800
灌　南　Guannan	31 808	309 100
响　水　Xiangshui	30 072	206 600
滨　海　Binhai	32 234	404 200
射　阳　Sheyang	61 084	342 100
东　台　Dongtai	59 640	469 600
大　丰　Dafeng	51 994	311 600
浙　江　Zhejiang	**1 810 052**	**7 994 500**
象　山　Xiangshan	197 947	281 000
宁　海　Ninghai	25 961	340 400
余　姚　Yuyao	62 883	460 600
慈　溪　Cixi	83 686	787 000
奉　化　Fenghua	59 792	243 100
洞　头　Dongtou	8 700	51 500
平　阳　Pingyang	70 900	410 600
苍　南　Cangnan	78 300	658 700
瑞　安　Rui'an	113 700	622 700
乐　清　Yueqing	238 100	663 400
海　盐　Haiyan	66 400	200 100
海　宁　Haining	123 861	315 000
平　湖　Pinghu	127 292	212 400
绍　兴　Shaoxing	177 574	448 000
上　虞　Shangyu	117 856	381 200

10-21 续表2 continued

沿海县 Coastal County	城镇单位在岗职工人数 Urban Employed Persons	乡村从业人员 Rural Laborer
岱　山　Daishan	10 700	84 500
嵊　泗　Shengsi	8 900	28 000
玉　环　Yuhuan	44 700	335 600
三　门　Sanmen	25 900	220 400
温　岭　Wenling	84 200	660 600
临　海　Linhai	82 700	589 700
福　建　Fujian	**1 410 271**	**5 879 893**
连　江　Lianjiang	37 573	292 376
罗　源　Luoyuan	14 027	93 168
平　潭　Pingtan	15 793	183 682
福　清　Fuqing	185 715	532 187
长　乐　Changle	39 369	259 919
仙　游　Xianyou	43 806	487 387
惠　安　Hui'an	178 178	461 340
石　狮　Shishi	74 284	113 229
晋　江　Jinjiang	478 773	645 877
南　安　Nan'an	85 985	774 284
云　霄　Yunxiao	21 555	163 392
漳　浦　Zhangpu	52 331	425 616
诏　安　Zhao'an	23 569	302 071
东　山　Dongshan	16 769	78 484
龙　海　Longhai	76 040	380 269
霞　浦　Xiapu	17 414	214 382
福　安　Fu'an	28 582	207 190
福　鼎　Fuding	20 508	265 040
山　东　Shandong	**1 082 088**	**5 420 451**
胶　州　Jiaozhou	124 490	340 195

10-21 续表3 continued

沿海县 Coastal County	城镇单位在岗职工人数 Urban Employed Persons	乡村从业人员 Rural Laborer
即墨 Jimo	129 481	553 647
胶南 Jiaonan	82 956	351 286
垦利 Kenli	23 608	77 621
利津 Lijin	16 012	138 654
广饶 Guangrao	60 471	254 257
长岛 Changdao	5 113	12 548
龙口 Longkou	69 947	258 287
莱阳 Laiyang	68 267	412 575
莱州 Laizhou	64 811	374 550
蓬莱 Penglai	39 139	204 961
招远 Zhaoyuan	60 318	219 699
寿光 Shouguang	62 890	402 953
海阳 Haiyang	29 605	344 268
昌邑 Changyi	28 774	263 533
文登 Wendeng	66 545	304 253
荣成 Rongcheng	75 041	188 762
乳山 Rushan	42 080	279 207
无棣 Wudi	17 940	250 001
沾化 Zhanhua	14 600	189 194
广东 Guangdong	**638 000**	**7 073 376**
南澳 Nan'ao	5 110	25 205
台山 Taishan	45 883	496 824
恩平 Enping	32 865	166 567
遂溪 Suixi	39 001	422 669
徐闻 Xuwen	36 175	305 197
廉江 Lianjiang	54 983	702 945
雷州 Leizhou	56 478	641 823
吴川 Wuchuan	34 012	513 214

10-21 续表4 continued

沿海县 Coastal County	城镇单位在岗职工人数 Urban Employed Persons	乡村从业人员 Rural Laborer
电　白　Dianbai	49 530	630 711
惠　东　Huidong	58 709	357 950
海　丰　Haifeng	28 865	383 885
陆　丰　Lufeng	47 019	571 669
阳　西　Yangxi	22 242	247 194
阳　东　Yangdong	28 454	249 705
饶　平　Raoping	29 440	394 302
揭　东　Jiedong	31 002	592 525
惠　来　Huilai	38 232	370 991
广　西　Guangxi	**44 743**	**384 500**
东　兴　Dongxing	7 079	52 200
合　浦　Hepu	37 664	332 300
海　南　Hainan	**188 528**	**1 867 861**
琼　海　Qionghai	18 467	181 159
儋　州　Danzhou	27 635	302 599
文　昌　Wenchang	23 959	219 671
万　宁　Wanning	21 105	178 597
东　方　Dongfang	18 895	172 450
澄　迈　Chengmai	22 526	188 375
临　高　Lingao	12 961	184 690
昌　江　Changjiang	17 098	80 007
乐　东　Ledong	16 201	224 880
陵　水　Lingshui	9 681	135 433

注：本表各省数据为合计数，缺少上海崇明县，福建金门县数据。

Note: The data for the provinces are the totals, and lacking the data for Chongming of Shanghai Province; Jinmen of Fujian Province.

主要统计指标解释

1. 国内(或地区)生产总值 指一个国家（或地区）所有常驻单位在一定时期内生产活动的最终成果。国内生产总值有三种表现形态，即价值形态、收入形态和产品形态。从价值形态看，它是所有常驻单位在一定时期内生产的全部货物和服务价值超过同期中间投入的全部非固定资产货物和服务价值的差额，即所有常驻单位的增加值之和；从收入形态看，它是所有常驻单位在一定时期内创造并分配给常驻单位和非常驻单位的初次收入分配之和；从产品形态看，它是所有常驻单位在一定时期内最终使用的货物和服务价值与货物和服务净出口价值之和。在实际核算中，国内生产总值有三种计算方法，即生产法、收入法和支出法。三种方法分别从不同的方面反映国内生产总值及其构成。

2. 三次产业 是根据社会生产活动历史发展的顺序对产业结构的划分，产品直接取自自然界的部门称为第一产业，对初级产品进行再加工的部门称为第二产业，为生产和消费提供各种服务的部门称为第三产业。它是世界上较为通用的产业结构分类，但各国的划分不尽一致。我国的三次产业划分是：

第一产业：是指农、林、牧、渔业。

第二产业：是指采矿业，制造业，电力、燃气及水的生产和供应业，建筑业。

第三产业：是指除第一、二产业以外的其他行业。第三产业包括：交通运输、仓储和邮政业，信息传输、计算机服务和软件业，批发和零售业，住宿和餐饮业，金融业，房地产业，租赁和商务服务业，科学研究、技术服务和地质勘查业，水利、环境和公共设施管理业，居民服务和其他服务业，教育，卫生、社会保障和社会福利业，文化、体育和娱乐业，公共管理和社会组织，国际组织。

3. 增加值 是指各行各业生产经营和劳务活动的最终成果，采用生产法和收入法两种方法计算。

生产法 是从货物和服务活动在生产过程中形成的总产品入手，剔除生产过程中投入的中间产品价值，得到新增价值的方法。

收入法 又称分配法。按收入法计算国内生产总值是从生产过程创造的收入的角度对常驻单位的生产活动成果进行核算；按照此法计算，增加值由劳动者报酬、固定资产折旧、生产税净额和营业盈余四个部分组成。

4. 年末总人口 是指每年 12 月 31 日 24 时一定地区范围内的有生命的个人的人口总和。

5. 财政收入 指国家财政参与社会产品分配所取得的收入，是实现国家职能的财力保证。财政收入所包括的内容几经变化，目前主要包括：

(1)各项税收：包括增值税、营业税、消费税、土地增值税、城市维护建设税、资源税、城市土地使用税、企业所得税、个人所得税、关税、证券交易印花税、车辆购置税、农牧业税和耕地占用税等。

(2)专项收入：包括排污费收入、城市水资源费收入、矿产资源补偿费收入、教育费附加收入等。

(3)其他收入：包括利息收入、基本建设贷款归还收入、基本建设收入、捐赠收入等。

(4)国有企业亏损补贴：此项为负收入，冲减财政收入。主要包括对工业企业、商业企业、粮食企业的补贴。

6. 财政支出 国家财政将筹集起来的资金进行分配使用，以满足经济建设和各项事业的需要。

7. 基本建设支出 指按国家有关规定，属于基本建设范围内的基本建设有偿使用、拨款、资本金支出以及经国家批准对专项和政策性基建投资贷款，在部门的基建投资额中统筹支付的贴息支出。

8. 普通高等学校 指按照国家规定的设置标准和审批程序批准举办的，通过全国普通高等学校统一招生考试，招收高中毕业生为主要培养对象，实施高等教育的全日制大学、独立设置的学院和高等专科学校、高等职业学校和其他机构。

大学、独立设置的学院主要实施本科层次以上教育，高等专科学校、高等职业学校实施专科层次教育，其他机构是承担国家普通招生计划任务不计校数的机构。包括普通高等学校分校和批准筹建的普通高等学校等。

9. 卫生机构 包括医疗机构、疾病预防控制中心(防疫站)、采供血机构、卫生监督及监测(检验)机构、医学科研和在职培训机构、健康教育所等。

10. 医疗机构 包括医院、社区卫生服务中心(站)、疗养院、卫生院、门诊部、诊所(卫生所、医务室)、妇幼保健院(所、站)、专科疾病防治院(所、站)、急救中心(站)和临床检验中心。医疗机构分为非赢利性医疗机构和赢利性医疗机构。

11. 单位国内生产总值能耗 指一定时期内，一个国家或地区每生产一个单位的国内生产总值所消耗的能源。计算公式为：

$$单位国内生产总值能源=\frac{能源消费总量}{国内生产总值}$$

12. 单位国内生产总值电耗 指一定时期内，一个国家或地区每生产一个单位的国内生产总值所消耗的电力。计算公式为：

$$单位国内生产总值电耗=\frac{全社会用电量}{国内生产总值}$$

13. 单位工业增加值能耗 指一定时期内，一个国家或地区每生产一个单位的工业增加值所消耗的能源。计算公式为：

$$单位工业增加值能耗=\frac{工业能源消费总量}{工业增加值}$$

14. 用水总量 指分配给各类用户的包括输水损失在内的毛用水量之和，不包括海水直接利用量。

15. 全社会固定资产投资 是以货币形式表现的在一定时期内全社会建造和购置固定资产的工作量以及与此有关的费用的总称。该指标是反映固定资产投资规模、结构和发展速度的综合性指标,又是观察工程进度和考核投资效果的重要依据。全社会固定资产投资按登记注册类型可分为国有、集体、个体、联营、股份制、外商、港澳台商、其他等。

16. 进出口总额 指实际进出我国国境的货物总金额。包括对外贸易实际进出口货物，来料加工装配进出口货物，国家间、联合国及国际组织无偿援助物资和赠送品，华侨、港澳台同胞和外籍华人捐赠品，租赁期满归承租人所有的租赁货物，进料加工进出口货物，边境地方贸易及边境地区小额贸易进出口货物(边民互市贸易除外)，中外合资企业、中外合作经营企业、外商独资经营企业进出口货物和公用物品，到、离岸价格在规定限额以上的进出口货样和广告品(无商业价值、无使用价值和免费提供出口的除外)，从保税仓库提取在中国境内销售的进口货物，以及其他进出口货物。该指标可以观察一个国家在对外贸易方面的总规模。我国规定出口货物按离岸价格统计，进口货物按到岸价格统计。

17. 商品经营单位所在地进、出口额 指在所在地海关注册登记的有进出口经营权的企业实际进、出口额。

18. 城镇家庭可支配收入 指家庭成员得到可用于最终消费支出和其它非义务性支出以及储蓄的总和，即居民家庭可以用来自由支配的收入。它是家庭总收入扣除交纳的所得税、个人交纳的社会保障支出以及记账补贴后的收入。计算公式为：

可支配收入=家庭总收入-交纳所得税-个人交纳的社会保障支出-记账补贴

19. 城镇家庭消费性支出 指家庭用于日常生活的支出，包括食品、衣着、家庭设备用品及服务、医疗保健、交通和通信、娱乐教育文化服务、居住、杂项商品和服务等八大类支出。

20. 总收入 指调查期内农村住户和住户成员从各种来源渠道得到的收入总和。按收入的性质划分为工资性收入、家庭经营收入、财产性收入和转移性收入。

21. 纯收入 指农村住户当年从各个来源得到的总收入相应地扣除所发生的费用后的收入总和。计算方法：

纯收入=总收入-税费支出-家庭经营费用支出-生产性固定资产折旧-赠送农村亲友支出

纯收入主要用于再生产投入和当年生活消费支出，也可用于储蓄和各种非义务性支出。“农民人均纯收入”按人口平均的纯收入水平，反映的是一个地区或一个农户农村居民的平均收入水平。

22. 人口数 指一定时点、一定地区范围内有生命的个人总和。

年度统计的年末人口数指每年 12 月 31 日 24 时的人口数。年度统计的全国人口总数内未包括香港、澳门特别行政区和台湾省以及海外华侨人数。

23. 城镇人口 城镇人口是指居住在城镇范围内的全部常住人口

24. 就业人员 指在 16 周岁及以上，从事一定社会劳动并取得劳动报酬或经营收入的人员。这一指标反映了一定时期内全部劳动力资源的实际利用情况，是研究我国基本国情国力的重要指标。

Explanatory Notes on Main Statistical Indicators

1. Gross Domestic Product (GDP) refers to the final result of the primary distribution of the income created by all the resident units of a country (or a region) during a certain period of time. Gross domestic product is expressed in three different forms, i.e. value, income, and products respectively. The form of value refers to the total value of all products and services produced by all resident units during a certain period of time minus total value of intermediate input of materials and services of the nature of non-fixed assets or the summation of the value added of all resident units: the form of income includes all the income created by all resident units and distributed primarily to all resident and non-resident units; the form of products refers to the summation of the value of the products and services finally used and the net export value of products and services by all resident units during a given period of time. In the practice of national accounting, gross domestic product is calculated with three approaches, i.e. production approach, income approach, and expenditure approach, which reflect the gross domestic product and its composition from different aspects.

2. Three Industries Industrial structure is classified according to the sequence of historical development of social productive activities. Primary industry refers to the extraction of natural resources; secondary industry involves processing of primary products; and tertiary industry provides services of various kinds for production and consumption. The above classification is universal in the world although it varies to some extent from country to country. The three industries in China are divided as follows:

Primary industry: refers to farming, forestry, sideline production and fishery.

Secondary industry: refers to such industries as mining, manufacturing, production and supply of electric power, fuel gas and water, and construction.

Tertiary industry: refers to all the other industries not included in the primary or secondary industries. It includes such industries as communications and transportation, storage and postal service; information transmission, computer service and software; wholesale and retailing; accommodation and catering; financial service; real estate; charter business and commercial affairs service; scientific research, technological service and geological survey; water conservancy, environmental and other public facilities management; residents service and other service trades; education, health, social security and social welfare; culture, sports and entertainment business as well as public administration and social organizations, and international organizations.

3. Added Value refers to the final result of production operation and labor activities of all trades and professions, which is calculated by using the methods of production and income.

Production Method: refers to the method whereby to get the newly added value by proceeding from the gross product of goods and service activities occurring in the course of production and then rejecting the value of intermediate product input in the course of production.

Income Method: is also called the distribution method. The calculation of the gross domestic product (GDP) by the income method is the accounting of the result of productive activities of permanent units from the angle of the income created in the course of production. According to this method, the added value is composed of the payment for laborers, depreciation for fixed assets, net tax on production and business surplus.

4. Total Population by the End of the Year refers to the sum of living individuals within a particular range of area at 24:00 on December 31 of each year.

5. Government Revenue refers to the income obtained by the government finance through participating in the distribution of social products. It is the financial guarantee to ensure government functioning. The contents of government revenue have changed several times. Now it includes the following main items:

(1) Various tax revenues, including value added tax, business tax, consumption tax, land

value-added tax, tax on city maintenance and construction, resources tax, tax on use of urban land, enterprise income tax, personal income tax, tariff, stamp tax on security transactions, tax on purchase of motor vehicles, tax on agriculture and animal husbandry and tax on occupancy of cultivated land, etc.

(2) Special revenues, including revenues from the fee on sewage treatment, fee on urban water resources, fee for the compensation of mineral resources and extra-charges for education, etc.

(3) Other revenues, including revenues from interest, repayment of capital construction loan, capital construction projects, and donations and grants.

(4) Subsidies for the losses of State-owned enterprises. This is an item of negative revenue, counteracting revenues and consisting of subsidies to industrial, commercial and grain purchasing and supply enterprises.

6. Government Expenditure refers to the distribution and use of the funds which the government finance has raised, so as to meet the needs of economic construction and various causes.

7. Expenditure for capital construction It refers to the non-gratuitous use of, appropriation of funds for and capital outlay on capital construction in the area of capital construction. It also covers the loans on capital construction approved by the government for special purposes or policy purposes and the expenditure with discount paid in an overall way within the amount of the funds appropriated to the departments for capital construction.

8. Regular Institutions of Higher Learning refer to educational establishments set up according to the government evaluation and approval procedures, enrolling graduates from senior secondary schools and providing higher education courses and training for senior professionals. They include full-time universities, colleges, institutions of higher professional education, institutions of higher vocational education and others.

Universities and colleges primarily provide undergraduate courses; institutions of higher professional education and institutions of higher vocational education primarily provide professional trainings; and others refer to educational establishments, which are responsible for enrolling higher education students under the State Plan but not enumerated in the total number of schools, including: branch schools of universities and colleges, and universities and colleges that have been approved and under plan for construction.

9. Health Care Institutions include: medical institutions, disease prevention and control centres (epidemic prevention stations), blood gathering and supplying institutions, health supervision and inspection (check up) institutions, medicinal scientific research and on-job training institutions, health education centres and so on.

10. Medical Organizations include: hospitals, health service centres (stations) in communities, sanatoria, health centres, out-patient clinics, clinics (health stations and infirmaries), maternity and child care agencies (centres and stations), special disease prevention and curing agencies (centres and stations), first aid centres (stations) and clinical inspection centres. Medical organizations are grouped by two types: profit-making and non-profit-making medical organizations.

11. Energy Consumption per Unit of GDP refers to the energy consumption per unit of Gross Domestic Product in a country or the Gross Regional Product in a region in the same reference period. The formula is:

$$\text{Energy Consumption per Unit of GDP} = \frac{\text{Total Energy Consumption}}{\text{Gross Domestic Product}}$$

12. Electricity Consumption per Unit of GDP refers to the electricity consumption per unit of Gross Domestic Product in a country or the Gross Regional Product in a region in the same reference period. The formula is:

$$\text{Electricity Consumption per Unit of GDP} = \frac{\text{Total Electricity Consumption}}{\text{Gross Domestic Product}}$$

13. Energy Consumption per Unit of Industrial Value-added refers to the energy consumption per unit of industrial value-added in a country or region in the same reference period. The formula is:

$$\text{Energy Consumption per Unit of Industrial Value-added} = \frac{\text{Industry Energy Consumption}}{\text{Industrial Value-added.}}$$

14. Gross Amount of Water Used refers to gross water use distributed to users, including loss during transportation, broken down into use by agriculture, industry, living consumption and ecological protection.

15. Total Investment in Fixed Assets in the Whole Country refers to the volume of activities in construction and purchases of fixed assets of the whole country and related fees, expressed in monetary terms during the reference period. It is a comprehensive indicator which shows the size, structure and growth of the investment in fixed assets, providing a basis for observing the progress of construction projects and evaluating results of investment. Total investment in fixed assets in the whole country includes, by type of ownership, the investment by State-owned units, collective-owned units, individuals, joint ownership units, share-holding units, as well as investments by entrepreneurs from foreign countries and from Hong Kong, Macao and Taiwan, and by other units.

16. Total Imports and Exports at Customs refer to the real value of commodities imported and exported across the border of China. They include the actual imports and exports through foreign trade, imported and exported goods under the processing and assembling trades and materials, supplies and gifts as aid given gratis between governments and by the United Nations and other international organizations, and contributions donated by overseas Chinese compatriots in Hong Kong and Macao and Chinese with foreign citizenship, leasing commodities owned by tenant at the expiration of leasing period, the imported and exported commodities processed with imported materials, commodities trading in border areas (excluding mutual exchange goods), the imported and exported commodities and articles for public use of the Sino-foreign joint ventures, cooperative enterprises and ventures with sole foreign investment. Also included is the import or export of samples and advertising goods for which the CIF or FOB value is beyond the permitted ceiling (excluding goods of no trading or use value and free commodities for export), imported goods sold in China from bonded warehouses and other imported or exported goods. The indicator of the total imports and exports at customs can be used to observe the total size of external trade in a country. In accordance with the stipulation of the Chinese government, imports are calculated at CIF, while exports are calculated at FOB.

17. Import-Export Value by Location of China's Foreign Trade Managing Units refers to actual value of imports and exports carried out by corporations which have been registered by the local customs house and are vested with right to run import export business.

18. Disposable Income of Urban Households refers to the actual income at the disposal of members of the households which can be used for final consumption, other non-compulsory expenditure and savings. This equals to total income minus income tax, personal contribution to social security and subsidy for keeping diaries in being a sample household. The following formula is used:

Disposable income = total household income − income tax − personal contribution to social security - subsidy for keeping diaries for a sampled household

19. Consumption Expenditure of Urban Households refers to total expenditure of households for consumption in daily life, including expenditure on the eight categories of food; clothing; household appliances and services; health care and medical services; transport and communications; recreation, education and cultural services; housing; and miscellaneous goods and services.

20. Total Income refers to the sum of income earned from various sources by the rural households and their members during the reference period, and is classified as income from wages and salaries, household operations, properties and transfers.

21. Net Income refers to the total income of rural households from all sources minus all corresponding expenses. The formula for calculation is as follows:

Net income = total income − taxes and fees paid − household operation expenses − taxes and fees − depreciation of fixed assets for production − gifts to non-rural relatives

Net income is mainly used as input for reinvestment in production and as consumption expenditure of the year, and also used for savings and non-compulsory expenses of various forms. "Per capita net income of farmers" is the level of net income averaged by population, reflecting the average income level of rural households in a given area.

22. Total Population refers to the total number of people alive at a certain point of time within a given area.

The annual statistics on total population is taken at midnight, the 3lst of December, not including residents in Taiwan province, Hong Kong and Macao and overseas Chinese.

23. Urban Population refers to all people residing in cities and towns.

24. Employed Persons refers to persons aged 16 and over who are engaged in gainful employment and thus receive remuneration payment or earn business income. This indicator reflects the actual utilization of total labour force during a certain period of time and is often used for the research on China’s economic situation and national power.

11

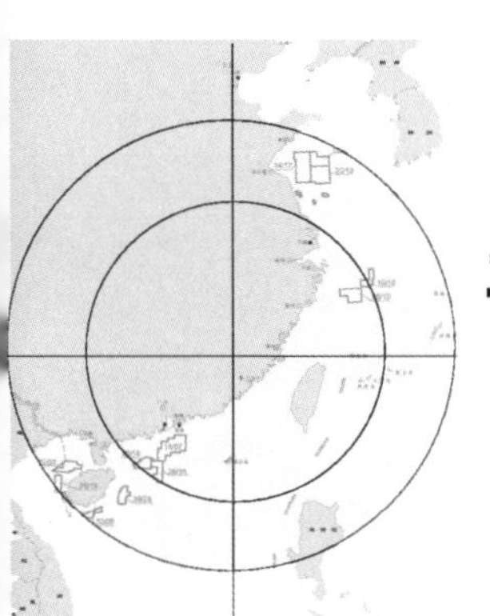

部分世界海洋经济统计资料

Part of the World's Marine Economic Statistics Data

11-1 世界海洋面积
The World Ocean Area

区 域 Region	海洋面积（平方公里） Ocean Area (km^2)	占世界海洋面积的比重（%） Proportion in the World Ocean Area (%)	占地球表面面积的比重（%） Proportion in the Earth's Surface Area (%)
合 计 Total	**361 000 000**	**100.0**	**70.8**
太 平 洋 Pacific Ocean	178 334 000	49.4	35.0
大 西 洋 Atlantic Ocean	91 694 000	25.4	18.0
印 度 洋 Indian Ocean	76 171 000	21.1	14.9
北 冰 洋 Arctic Ocean	14 801 000	4.1	2.9

资料来源：《2010国际统计年鉴》。
Source: *International Statistical Yearbook 2010*.

11-2 主要沿海国家（地区）海岸线长度
Length of Coastline of Major Coastal Countries (Regions)

单位：公里 (km)

国家和地区 Country and Region	海岸线长度 Length of Coastline
美 国 United States	22 680
日 本 Japan	30 000
德 国 Germany	1 300
英 国 United Kingdom	11 450
法 国 France	3 000
意大利 Italy	7 000
加拿大 Canada	20 000
澳大利亚 Australia	20 125
俄罗斯联邦 Russian Fed.	34 000
波 兰 Poland	491
印 度 India	6 083
印度尼西亚 Indonesia	35 000
菲律宾 Philippines	18 533
泰 国 Thailand	3 219
马来西亚 Malaysia	4 675
新加坡 Singapore	193
缅 甸 Myanmar	3 060
孟加拉国 Bangladesh	580
土耳其 Turkey	7 200
韩 国 Republic of Korea	2 413
埃 及 Egypt	2 450
墨西哥 Mexico	9 330
巴 西 Brazil	7 400
阿根廷 Argentina	4 989

11-3 世界主要沿海国家（地区）土地和人口（2008年）

Land Area and Population of Major Coastal Countries (Regions), 2008

国家和地区 Country and Region	国土面积(万平方公里) Area of Territory (10 000 km^2)	年中人口（万人） Mid-year Population (10 000 persons)	人口密度（人/平方公里） Population Density (people per km^2)
世界总计 World Total	**13 409.7***	**669 725**	**52**
中 国 China	960.0	132 466	142
美 国 United States	963.2	30 406	33
日 本 Japan	37.8	12 770	350
加拿大 Canada	998.5	3 331	4
德 国 Germany	35.7	8 211	235
英 国 United Kingdom	24.4	6 141	254
法 国 France	54.9	6 228	114
意大利 Italy	30.1	5 983	203
墨西哥 Mexico	196.4	10 635	55
印度尼西亚 Indonesia	190.5	22 735	125
马来西亚 Malaysia	33.0	2 701	82
泰 国 Thailand	51.3	6 739	132
新加坡 Singapore	0.1	484	6 943
韩 国 Republic of Korea	10.0	4 861	502
印 度 India	328.7	113 996	383
巴 西 Brazil	851.5	19 197	23
俄罗斯联邦 Russian Fed.	1 709.8	14 195	9

注：*数据来源于《2010年中国统计年鉴》。

Note: *The data come from the China Statistical Yearbook 2010.

11-4 主要沿海国家（地区）国内生产总值
Gross Domestic Product of Major Coastal Countries (Regions)

国家和地区 Country and Region	2008年			2009年
	国内生产总值（亿美元）GDP (100 million US$)	人均国民总收入（美元）GNI per Captia (current US$)	国内生产总值增长率(%) Growth Rate of GDP(%)	国内生产总值（亿美元）GDP (100 million US$)
美 国 United States	142 043	47 580	0.4	142 563
日 本 Japan	49 093	38 210	-0.7	50 681
德 国 Germany	36 528	42 440	1.3	33 527
英 国 United Kingdom	26 456	45 390	0.7	21 836
法 国 France	28 531 *	42 250 *	0.3	26 759
意大利 Italy	22 930	35 240	-1.0	21 183
加拿大 Canada	14 001	41 730	0.4	13 364
俄罗斯联邦 Russian Fed.	16 078	9 620	5.6	12 292
印 度 India	12 175	1 070	7.4	12 360
印度尼西亚 Indonesia	5 144	2 010	6.1	5 394
泰 国 Thailand	2 607	2 840	2.6	2 639
马来西亚 Malaysia	1 949	6 970	4.6	1 915
新加坡 Singapore	1 819	34 760	1.2	1 771
韩 国 Republic of Korea	9 291	21 530	2.2	8 325
墨西哥 Mexico	10 860	9 980	1.4	8 749
巴 西 Brazil	16 125	7 350	5.1	15 740

注：*包括法属圭亚那、瓜德罗普、马提尼克和留尼汪。

Note: *Including French Guiana, Guadeloupe, Martinique and Reunion.

11-5 世界主要沿海国家（地区）国内生产总值产业构成(2008年)
Industrial Composition of GDP of Major Coastal Countries (Regions), 2008

国家和地区 Country and Region	国内生产总值产业构成（%） Industry Structure of GDP(%)		
	第一产业 Primary Industy	第二产业 Secondary Industry	第三产业 Tertiary Industy
世　界 World	**3.0**	**28.0**	**69.0**
美　国 United States			
日　本 Japan			
德　国 Germany			
英　国 United Kingdom			
法　国 France			
意大利 Italy			
墨西哥 Mexico	3.8	37.1	59.1
印度尼西亚 Indonesia	14.4	48.1	37.5
马来西亚 Malaysia			
泰　国* Thailand	10.8	43.8	45.3
新加坡 Singapore	0.1	27.8	72.2
韩　国 Republic of Korea	2.6	37.1	60.3
印　度 India	17.6	29.0	53.4
巴　西 Brazil	6.7	28.0	65.3
俄罗斯联邦* Russian Fed.	4.8	38.6	56.7

注：*为2006年数据。
Note: *The data for 2006.

11-6 世界主要沿海国家（地区）就业人数
Employment in the Major Coastal Countries (Regions)

单位：万人 (10 000 persons)

国家和地区 Country and Region	2000	2004	2005	2006	2007	2008
中　　国[①] China	72 085	75 200	75 825	76 400	76 990	77 480
中国香港[②] Hong Kong,China	321	327	334	340	348	352
孟加拉国 Bangladesh	5 176		4 736			
印　　度 India	36 897					
印度尼西亚 Indonesia	8 984	9 372	9 396	9 546	9 993	10 255
以 色 列[④] Israel	222	240	249	257	268	278
日　　本 Japan	6 446	6 329	6 356	6 382	6 412	6 385
哈萨克斯坦 Kazakhstan	620	718	726	740	763	786
韩　　国[④] Republic of Korea	2 116	2 256	2 286	2 315	2 343	2 358
马来西亚[④⑤] Malaysia	927	998	1 005	1 028	1 054	1 066
巴基斯坦[③④]Pakistan	3 685	4 201	4 292	4 695	4 765	
菲 律 宾[④] Philippines	2 745	3 161	3 231	3 264	3 356	3 409
新 加 坡 Singapore	148	163		180	180	185
斯里兰卡[⑦] Sri Lanka	631[③]	739[⑧]	752	711	704	717
泰　　国[④] Thailand	3 300[⑨]	3 571	3 630	3 634	3 712	3 784
土 耳 其[④] Turkey	2 158	2 179	2 205	2 233	2 074	2 119
越　　南 Viet Nam	3 837	4 232	4 253	4 334	4 417	4 492
埃　　及[④⑤] Egypt	1 720	1 872	1 934	2 044	2 172	

11-6 续表 continued

国家和地区 Country and Region	2000	2004	2005	2006	2007	2008
南　　非 South Africa	1 224	1 164	1 230	1 280	1 323	1 371
加 拿 大[④⑩] Canada	1 476	1 595	1 617	1 648	1 687	1 713
墨 西 哥[⑪] Mexico	3 804	4 056	4 079	4 220	4 291	4 387
美　　国[④⑥] United States	13 521	13 925	14 173	14 443	14 605	14 536
阿 根 廷[③⑫] Argentina	826	942	964	1 004	1 012	1 028
巴　　西[③] Brazil	6 563	8 460	8 719	8 932	9 079	
委内瑞拉[④] Venezuela	896[③]	1042[③]	1073[③]	1112[③]	1 149	1 186
法　　国 France	2 326	2 480	2 498	2 513	2 557	2 591
德　　国 Germany	3 660	3 566	3 657	3 732	3 816	3 873
意 大 利 Italy	2 123	2 240	2 256	2 299	2 322	2 340
荷　　兰 Netherlands	780	793	796	811	831	846
波　　兰[④] Poland	1 453	1 380	1 412	1 459	1 524	1 580
俄罗斯联邦[⑬] Russian Fed.	6 507	6 728	6 817	6 886	7 057	7 097
西 班 牙[④⑥] Spain	1 551	1 797	1 897	1 975	2 036	2 026
乌 克 兰[⑭] Ukraine	2 018	2 030	2 068	2 073	2 090	2 097
英　　国[⑥] United Kingdom	2 740	2 837	2 867	2 893	2 910	2 948
澳大利亚[④] Australia	895	962	997	1 022	1 051	1 074
新 西 兰[④] New Zealand	180	202	208	213	217	219

注：①不包括军人和返聘的退休人员。②不包括军人和域外机构人员。③10岁及以上。④不包括军人。⑤15岁至64岁。⑥16岁及以上。⑦不包括北部和东部省。⑧不包括北部省。⑨13岁及以上。⑩不包括生活在保护的居民。⑪14岁以上。⑫部分城市地区。⑬15岁至72岁。⑭15岁至70岁。

Note: ①Excluding armed forces and reemployed retired persons. ②Excluding marine, military and institutional populations. ③Persons aged 10 years. ④Excluding armed forces. ⑤Persons aged 15 to 64 years. ⑥Persons aged 16 years and over. ⑦Excluding Northern and Eastern provinces. ⑧Excluding Northern provinces. ⑨Persons aged 13 years and over.⑩Excluding residents of the territories and indigenous persons living on reserves. ⑪Person aged 14 years and over. ⑫Some urban agglomerations. ⑬Persons aged 15 to 72 years. ⑭Persons aged 15 to 70 years.

11-7 主要沿海国家（地区）渔业增加值
Fishery Added Value of Major Coastal Countries (Regions)

单位：亿本币 (100 million local currency units)

国家和地区 Country and Region	渔业增加值 Fishery Added Value				
	2000	2004	2005	2006	2007
中国 China	14 945	21 413	23 070	24 737	
中国香港 Hong Kong, China	9	9	8	8	
孟加拉国 Bangladesh	1 367	1 478	1 546	1 632	1 733
印度 India	2 147	2 796	3 324		
印度尼西亚* Indonesia	30	53	59	73	
伊朗* Iran	2	3	4		
日本* Japan	9	8	8	7	
哈萨克斯坦 Kazakhstan	40	58	60	73	86
韩国 Republic of Korea	21 547	18 110	17 226	15 599	14 832
马来西亚 Malaysia	51	53	56	66	69
菲律宾 Philippines	785	1 105	1 163	1 298	
新加坡 Singapore	2	2	2	2	2
泰国 Thailand	1 178	1 070	1 094	1 137	
越南* Viet Nam	15	27	33	38	46
加拿大 Canada	14	14			
墨西哥 Mexico	64	84			
美国 United States	980	1 422	1 288	1 254	
阿根廷 Argentina	6	13	16	20	
委内瑞拉* Venezuela	…	1	1		
法国 France	14	15	15	13	15
德国 Germany	2	2	2	2	
意大利 Italy	13	15	15	16	
荷兰 Netherlands	3	2	2	2	
波兰 Poland	2	2	2	2	
西班牙 Spain	15	17	16	16	
乌克兰 Ukraine	2	1	2	2	
英国 United Kingdom	4	4	4		
澳大利亚 Australia	249	268	275	223	
新西兰 New Zealand	4				

注：*万亿本币。

Note: *Trillion local currency units.

11-8 主要沿海国家（地区）旅馆和饭店业增加值
Added Value of Hotel Industry in the Major Coastal Countries (Regions)

单位：亿本币 (100 million local currency units)

国家和地区 Country and Region	旅馆和饭店业增加值 Added Value of Hotel Industry				
	2000	2004	2005	2006	2007
中国 China	2 146	3 665	4 193	4 833	
中国香港 Hong Kong, China	366	331	365		
孟加拉国 Bangladesh	146	220	251	285	327
印度 India	2 513	4 127	4 958		
伊朗* Iran	2	6	7		
哈萨克斯坦 Kazakhstan	148	516	681	842	1 092
韩国 Republic of Korea	148 618	184 206	188 713	197 262	205 598
马来西亚 Malaysia	80	102	112	124	144
菲律宾 Philippines	630	894	959	1 054	
新加坡 Singapore	34	33	37	41	46
泰国 Thailand	2 752	3 342	3 469	3 864	
越南* Viet Nam	14	23	29	36	45
埃及 Egypt	66	127	177	247	
加拿大 Canada	234	271			
墨西哥 Mexico	2 466	3 067			
美国 United States	2 614	3 137	3 310	3 536	
阿根廷 Argentina	78	99	126	163	
委内瑞拉* Venezuela	1	3	6		
法国 France	302	346	367	380	395
德国 Germany	301	318	328	338	
意大利 Italy	416	464	484	504	
荷兰 Netherlands	76	83	83	86	
波兰 Poland	83	95	107	111	
西班牙 Spain	434	574	607	641	
乌克兰 Ukraine	8	23	24	54	
英国 United Kingdom	260	331	327		
澳大利亚 Australia	158	194	201	210	
新西兰 New Zealand	21				

注：*万亿本币。

Note: *Trillion local currency units.

11-9 主要沿海国家（地区）运输、仓储和通讯业增加值

Added Values of Transportation,Storage and Communications Industries in the Major Coastal Countries (Regions)

单位：亿本币 (100 million local currency units)

国家和地区 Country and Region	运输、仓储和通讯业增加值 Added Values of Transportation, Storage and Commnuications Industries				
	2000	2004	2005	2006	2007
中国 China	6 161	9 304	10 836	12 032	
中国香港 Hong Kong,China	1 190	1 268	1 351	1 371	
孟加拉国 Bangladesh	1 974	3 444	3 829	4 321	4 843
印度 India	15 030	24 801	28 226		
印度尼西亚* Indonesia	65	142	181	231	
伊朗* Iran	48	98	116		
以色列 Israel	342	398	416	443	
日本* Japan	35	34	34	33	
哈萨克斯坦 Kazakhstan	2 985	6 912	8 968	11 788	14 221
韩国 Republic of Korea	361 387	509 690	524 295	538 143	574 511
马来西亚 Malaysia	249	333	360	386	423
巴基斯坦 Pakistan	4 010	6 756	7 597	9 332	10 566
菲律宾 Philippines	1 990	3 674	4 139	4 462	
新加坡 Singapore	208	256	272	285	312
斯里兰卡 Sri Lanka	1 359	2 426	2 906	3 495	
泰国 Thailand	3 959	4 925	5 197	5 671	
越南* Viet Nam	17	30	37	44	51
埃及 Egypt	271	468	612	625	
南非 South Africa	809	1 222	1 341	1 458	1 595
加拿大 Canada	702	838			
墨西哥 Mexico	5 568	7 255			
美国 United States	6 363	7 136	7 464	7 902	
阿根廷 Argentina	241	374	444	536	
巴西 Brazil	866	1 425	1 647		
委内瑞拉* Venezuela	5	12	16		
法国 France	776	965	998	1 020	1 080
德国 Germany	1 018	1 174	1 156	1 212	
意大利 Italy	777	960	981	1 009	
荷兰 Netherlands	266	323	327	338	
波兰 Poland	434	608	627	673	
俄罗斯联邦 Russian Fed.	5 929	16 620	19 251	22 822	26 696
西班牙 Spain	418	546	560	583	
乌克兰 Ukraine	198	427	474	561	686
英国 United Kingdom	693	793	811		
澳大利亚 Australia	507	652	671	743	
新西兰 New Zealand	82				

注：*万亿本币。

Note: *Trillion local currency units.

11-10 主要沿海国家（地区）海洋渔区标称渔获量
Nominal Catches of Marine Fishing Zone in the Major Coastal Countries (Regions)

国家和地区 Country and Region	2006	2007
世界总计 World Total	**80 068 385**	**80 029 324**
阿尔巴尼亚 Albania	3 287	2 899
阿尔及利亚 Algeria	145 762	148 436
安哥拉 Angola	214 949	297 440
安提瓜和巴布达 Antigua Barb	3 092	3 092
阿根廷 Argentina	1 138 363	962 638
澳大利亚 Australia	196 773	183 756
巴哈马 Bahamas	10 598	3 749
巴林 Bahrain	15 594	15 012
孟加拉国 Bangladesh	479 810	487 438
巴巴多斯 Barbados	1 974	1 800 *
比利时 Belgium	22 523	24 029
伯利兹 Belize	4 217	6 682
贝宁 Benin	9 821	7 701
巴西 Brazil	527 872	539 967
文莱 Brunei Darsm	1 992	2 241
保加利亚 Bulgaria	5 628	7 829
柬埔寨 Cambodia	60 500	60 000 *
喀麦隆 Cameroon	62 232	64 232
加拿大 Canada	1 036 743	973 663
佛得角群岛 Cape Verde	24 589	18 328
智利 Chile	4 160 848	3 806 085
哥伦比亚 Colombia	75 000 *	76 000 *
科摩罗群岛 Comoros	15 070 *	16 000
刚果 Congo Rep.	28 485	29 821
库克群岛 Cook Is.	3 700	3 200
哥斯达黎加 Costa Rica	21 000 *	20 735
科特迪瓦 Côte dIvoire	26 145	26 100 *
古巴 Cuba	26 118	33 572
塞浦路斯 Cyprus	2 135	2 426
丹麦 Denmark	867 812	652 982
多米尼加共和国 Dominican Rp.	11 045	12 228
厄瓜多尔 Ecuador	448 595	383 495
埃及 Egypt	119 606	130 748
萨尔瓦多 El Salvador	41 184	46 217

11-10 续表1 continued

国家和地区 Country and Region	2006	2007
赤道几内亚 Eq. Guinea	2 700 *	2 883
爱沙尼亚 Estonia	83 549	95 171
法罗群岛 Faeroe Is.	623 122	582 134
福克兰群岛 Falkland Is.	75 287	72 146
斐济 Fiji	43 002	46 489
芬兰 Finland	112 932	128 169
法国 France	572 333	510 176
波利尼西亚 Fr. Polynesia	13 356	13 027
加蓬 Gabon	32 162	29 500 *
冈比亚 Gambia	32 412	38 709
格鲁吉亚 Georgia	9 659 *	18 147
德国 Germany	276 395	227 301
加纳 Ghana	291 919	245 725
希腊 Greece	96 825	94 655
格陵兰岛 Greenland	247 011 *	247 011 *
格林纳达 Grenada	2 169	2 407
瓜德罗普 Guadeloupe	10 100	10 100
危地马拉 Guatemala	16 307	15 227
几内亚 Guinea	96 000 *	96 000 *
几内亚比绍 Guinea Bissau	6 050 *	6 050 *
圭亚那 Guyana	52 200 *	46 640
海地 Haiti	9 700 *	9 700 *
洪都拉斯 Honduras	16 794 *	12 778 *
冰岛 Iceland	1 326 924	1 398 982
印度 India	2 952 198	3 083 973
印度尼西亚 Indonesia	4 519 446	4 641 929
伊朗 Iran	374 447	329 571
伊拉克 Iraq	12 959	12 319
爱尔兰 Ireland	210 667	226 967
以色列 Israel	2 220	2 220 *
意大利 Italy	311 521	282 701
牙买加 Jamaica	17 430 *	16 148
日本 Japan	4 302 811	4 172 213
肯尼亚 Kenya	6 955	7 448
基里巴斯 Kiribati	23 600 *	21 598 *
朝鲜 Korea D. P. Rp.	200 000 *	200 000 *
韩国 Korea Rep.	1 768 990	1 852 403

11-10 续表2 continued

国家和地区 Country and Region	2006	2007
拉脱维亚 Latvia	140 061	154 966
黎巴嫩 Lebanon	3 541	3 541
利比里亚 Liberia	6 494	12 745
利比亚 Libya	34 647 *	31 921
立陶宛 Lithuania	153 111	185 639
马达加斯加 Madagascar	101 092	115 148
马来西亚 Malaysia	1 292 170	1 381 423
马尔代夫 Maldives	184 158	143 597
马耳他 Malta	1 330	1 235
马尔提尼克 Martinique	6 300	6 300
毛里塔尼亚 Mauritania	150 312	186 588 *
毛里求斯 Mauritius	8 681	7 906
墨西哥 Mexico	1 241 098	1 225 000 *
密克罗尼西亚 Micronesia	12 319	16 985
摩洛哥 Morocco	1 *	1 *
莫桑比克 Mozambique	75 882	68 189
缅甸 Myanmar	1 375 670	1 517 940
纳米比亚 Namibia	506 595	412 718
荷兰 Netherlands	433 235	411 602
荷属安的列斯群岛 Neth Antilles	6 247	3 662
新喀里多尼亚 New Caledonia	3 151	3 510
新西兰 New Zealand	476 019	488 143
尼加拉瓜 Nicaragua	35 221	24 598
尼日利亚 Nigeria	328 928	303 313
挪威 Norway	2 255 734	2 378 099
阿曼 Oman	147 669	151 744
巴基斯坦 Pakistan	349 421	340 190
帛琉群岛 Palau Is.	967	985
巴拿马 Panama	223 270	203 490
巴布亚新几内亚 Papua New Guinea	266 637	250 460
秘鲁 Peru	6 980 025	7 167 911
菲律宾 Philippines	2 159 130	2 333 175
波兰 Poland	126 296	132 761
葡萄牙 Portugal	229 076	252 801
波多黎各 Puerto Rico	2 042	1 675
卡塔尔 Qatar	16 376	15 190
留尼汪岛 Reunion	3 546	3 924

11-10 续表3 continued

国家和地区 Country and Region	2006	2007
俄罗斯 Russian Fed.	3 074 909	3 225 631
萨摩亚群岛 Samoa Is.	3 749	4 605
圣多美和普林西比 Sao Tome & Principe	4 150 *	4 150 *
沙特阿拉伯 Saudi Arabia	65 471	70 000 *
塞内加尔 Senegal	328 927	371 317
塞舌尔 Seychelles	93 121	65 871
塞拉利昂 Sierra Leone	134 146	130 535
新加坡 Singapore	3 103	3 522
索罗门群岛 Solomon Is.	39 336 *	31 272 *
索马里 Somalia	29 800 *	29 800 *
南非 South Africa	617 716	669 671
西班牙 Spain	944 714	802 682
斯里兰卡 Sri Lanka	238 735	271 375
苏丹 Sudan	5 000 *	5 699
苏里南 Suriname	30 421	29 277
瑞典 Sweden	267 607	236 707
叙利亚 Syria	3 395	3 381
坦桑尼亚 Tanzania	41 670	44 471
泰国 Thailand	2 484 803	2 244 944
多哥 Togo	19 879	14 905
汤加 Tonga	2 500 *	2 545 *
特立尼达和多巴哥 Trinidad & Tobago	8 445	8 406
突尼斯 Tunisia	110 024	102 079
土耳其 Turkey	488 966	589 129
特克斯和凯科斯群岛 Turks & Caicos	6 018	4 830
乌克兰 Ukraine	231 563	200 905
阿拉伯联合酋长国 United Arab Em.	87 000 *	87 000 *
英国 United Kingdom	621 402	617 174
美国 United States	4 833 547	4 742 326
乌拉圭 Uruguay	131 818	107 520
瓦努阿图 Vanuatu	88 085	85 356
委内瑞拉 Venezuela	410 910 *	405 910 *
越南 Viet Nam	1 824 800	1 977 400
也门 Yemen	229 660	179 916

注:*为联合国粮农组织估算值 (表11-11同)。
资料来源:《渔业统计年鉴》，联合国粮农组织，2007年Vol.100 (表11-11同)。
Note: * It is estimated by FAO from available sources of information or calculation.(Same as Table 11-11).
Source: *Yearbook of Fishery*, FAO, Vol. 100, 2007 (Same as Table 11-11).

11-11 主要沿海国家（地区）海藻和其他养殖品种捕捞产量
Capture Production of Seaweeds and Other Aquatic Plants in the Major Coastal Countries (Regions)

单位：吨 (ton)

国家和地区 Country and Region	2005	2006	2007
世界总计 World Total	**1 220 705**	**1 067 384**	**1 104 948**
阿根廷 Argentina	...	...	...
澳大利亚 Australia	14 167	15 504	2 223
加拿大 Canada	20 756	11 313	11 490
智利 Chile	409 851	305 748	336 472
爱沙尼亚 Estonia	809	394	1 608
斐济 Fiji	350 *	350 *	350 *
法国 France	23 054	19 160	37 736
冰岛 Iceland	21 096	20 964	21 867
印度尼西亚 Indonesia	7 730	4 996	5 230
意大利 Italy	1 600 *	1 400 *	1 400 *
日本 Japan	104 893	113 665	103 600
韩国 Korea Rep.	15 212	13 754	18 189
马达加斯加 Madagascar	5 225	5 300	3 650
墨西哥 Mexico	5 277	4 532	4 500 *
摩洛哥 Morocco	12 812	14 870	12 373
纳米比亚 Namibia	...	...	...
挪威 Norway	153 906	145 429	134 671
秘鲁 Peru	5 000	3 434	10 786
菲律宾 Philippines	299	314	351
葡萄牙 Portugal	624	765	495
俄罗斯 Russian Fed.	50 262	65 554	28 294
塞内加尔 Senegal			
南非 South Africa	6 619	6 600 *	6 600 *
西班牙 Spain	441	485	109
坦桑尼亚 Tanzania	52	278	214
乌克兰 Ukraine	522	1 121	1 892
美国 United States	68 065	6 362	2 272

11-12 主要沿海国家（地区）油气可开采储量（2006年）
Recoverable Reserves of Oil and Gas in the Major Coastal Countries (Regions), 2006

国家和地区 Country and Region	原油和液化天然气（万吨） Crude Oil and liquified Natural Gas (10 000 tons)	天然气（亿立方米） Natural Gas (100 million m^3)	油页岩(万吨) Oil Shale(10 000 tons)
中国 China	221 200	235 000	229 000
孟加拉国 Bangladesh	300	4 360	
印度 India	78 600	11 010	
印度尼西亚 Indonesia	57 000	27 540	
伊朗 Iran	1 734 000	267 400	
以色列 Israel	100	340	55 000
日本 Japan	900	510	
马来西亚 Malaysia	36 500	24 800	
缅甸 Myanmar	700	4 850	28 600
巴基斯坦 Pakistan	4 000	8 070	
菲律宾 Philippines	500	1 000	
泰国 Thailand	5 100	3 040	91 600
土耳其 Turkey	16 500	150	28 400
越南 Viet Nam	41 300	3 650	
埃及 Egypt	49 500	18 940	81 600
尼日利亚 Nigeria	482 300	51 500	
南非① South Africa	300	310	1 900
加拿大 Canada	210 600	16 330	219 200
墨西哥 Mexico	184 700	4 120	
美国 United States	369 100	58 660	30 156 600
阿根廷 Argentina	30 000	4 390	5 700
巴西 Brazil	159 100	3 060	1 173 400
委内瑞拉 Venezuela	1 126 900	43 150	
法国② France	1 700	100	100 200
德国 Germany	2 800	1 780	28 600
意大利③ Italy	10 600	1 700	1 044 600
荷兰 Netherlands	1 100	12 560	
波兰 Poland	1 600	750	700
俄罗斯联邦 Russian Fed.	1 002 700	478 200	3 547 000
西班牙 Spain	2 100	30	4 000
英国 United Kingdom	51 600	4 810	50 100
澳大利亚 Australia	22 500	7 550	453 100
新西兰 New Zealand	700	300	300

注：①包括博茨瓦纳、莱索托、斯威士兰和纳米比亚。②包括摩纳哥。③包括圣马力诺。
Note: ①Includes Botswana, Lesotho, Swaziland and Namibia. ②Includes Monaco.③Includes San Marino.

11-13 主要沿海国家（地区）油气产量（2008年）
Natural Gas and Crude Oil Production of the Major Coastal Countries (Regions) 2008

国家和地区 Country and Region	天然气产量（万亿焦耳） Production of Natural Gas (terajoule)	原油产量（万吨） Production of Crude Oil (10 000 tons)
中国 China	3 133 944	19 002
孟加拉国 Bangladesh	622 968	
印度 India	1 233 516	3 397
印度尼西亚 Indonesia	3 197 136	4 831
哈萨克斯坦 Kazakhstan	385 980	6 659
马来西亚 Malaysia	2 245 896	3 287
巴基斯坦 Pakistan	1 438 608	325
泰国 Thailand		1 140
加拿大 Canada	6 088 944	12 659
墨西哥 Mexico	3 233 544	14 552
美国 United States	22 335 420	33 505
阿根廷 Argentina	2 004 156	3 235
巴西 Brazil	934 716	9 215
德国 Germany	617 700	474
意大利 Italy	346 716	534
荷兰 Netherlands	2 524 716	
波兰 Poland	157 140	
俄罗斯联邦 Russian Fed.	22 351 968	48 756
乌克兰 Ukraine	703 176	418
英国 United Kingdom	2 913 780	6 539
澳大利亚 Australia	1 364 064	2 375
新西兰 New Zealand	170 544	265

11-14 主要沿海国家（地区）能源利用效率

Use Efficiency of Energy Resources in the Major Coastal Countries (Regions)

单位：吨标准油/万美元 (ton of oil equivalent per 10 000 US$)

国家和地区 Country and Region	单位国内生产总值能耗 Energy consumption per unit of GDP		
	2004	2005	2006
世界 World	**3.10**	**3.07**	**3.03**
中国 China	9.24	9.08	8.89
中国香港 Hong Kong,China	0.91	0.87	0.82
孟加拉国 Bangladesh	3.94	3.94	3.83
印度 India	8.81	8.35	8.00
印度尼西亚 Indonesia	8.75	8.47	8.16
伊朗 Iran	11.89	11.89	12.14
以色列 Israel	1.57	1.54	1.47
日本 Japan	1.09	1.06	1.03
哈萨克斯坦 Kazakhstan	18.85	18.92	18.52
韩国 Republic of Korea	3.32	3.20	3.10
马来西亚 Malaysia	4.96	5.58	5.46
巴基斯坦 Pakistan	8.48	8.08	7.91
菲律宾 Philippines	4.89	4.62	4.32
新加坡 Singapore	2.41	2.70	2.47
斯里兰卡 Sri Lanka	5.07	4.58	4.40
泰国 Thailand	6.50	6.40	6.24
土耳其 Turkey	2.66	2.65	2.63
越南 Viet Nam	12.15	11.46	10.79
埃及 Egypt	4.98	5.16	4.93
尼日利亚 Nigeria	16.99	16.97	15.98
南非 South Africa	8.45	7.95	7.67
加拿大 Canada	3.37	3.33	3.19
墨西哥 Mexico	2.68	2.77	2.66
美国 United States	2.19	2.14	2.06
阿根廷 Argentina	2.20	2.02	2.03
巴西 Brazil	2.94	2.93	2.91
委内瑞拉 Venezuela	4.73	4.55	4.24
法国 France	1.95	1.91	1.85
德国 Germany	1.80	1.77	1.73
意大利 Italy	1.61	1.62	1.58
荷兰 Netherlands	2.05	2.01	1.90
波兰 Poland	4.76	4.64	4.61
俄罗斯联邦 Russian Fed.	19.53	18.76	18.00
西班牙 Spain	2.16	2.13	2.04
乌克兰 Ukraine	32.75	31.67	28.32
英国 United Kingdom	1.45	1.43	1.37
澳大利亚 Australia	2.46	2.55	2.52
新西兰 New Zealand	2.95	2.82	2.82

11-15 主要沿海国家（地区）国际海运装货量和卸货量

International Ocean Shipping Loading and Unloading Capacities in the Major Coastal Countries (Regions)

单位：万吨 (10 000 tons)

国家和地区 Country and Region	国际海运装货量 International Ocean Shipping Loading Capacity		国际海运卸货量 International Ocean Shipping Unloading Capacity	
	2006	2007	2006	2007
中国 China	203 000 ①			
中国香港 Hong Kong,China	8 058	8 676	11 795	11 775
孟加拉国 Bangladesh	65	61	1 578	1 508
印度尼西亚 Indonesia	27 589	29 042	7 110	7 643
伊朗 Iran	3 132	3 286	5 714	6 492
以色列② Israel	1 321	1 479	1 700	2 872
韩国 Republic of Korea	21 992	23 835	45 494	48 042
马来西亚 Malaysia	7 563	8 081	8 780	9 757
巴基斯坦 Pakistan	938 ③		3 296 ③	
新加坡 Singapore	37 357 ①	40 301 ①		
埃及 Egypt	1 769 ③		3 534 ③	
美国 United States	33 778 ③		78 467 ③	
阿根廷④ Argentina		1 916		
法国 France	8 388 ③		18 925 ③	
德国 Germany	9 750	9 900	14 892	15 675
波兰 Poland	3 694	2 588	1 516	2 308
俄罗斯联邦 Russian Fed.	638		37	
西班牙 Spain	7 466		21 006	
乌克兰 Ukraine	5 699	5 322	1 269	1 528
澳大利亚 Australia	53 599	55 898	6 325	6 661
新西兰 New Zealand	1 859	1 951	1 543	1 539

注：①包括装货量和卸货量。②不包括石油。③2005年数字。④不包括转口。

Note: ①Including both goods loaded and unloaded. ②Excluding petroleum. ③Data for 2005. ④Excluding re-exports.

11-16 世界主要外贸海运量（2008年）
World Major Maritime Freight Traffic in Foreign Trade, 2008

品 种 Sort	海运量（百万吨） Freight Traffic (million tons)		海运周转量（十亿吨海里） Ton-Kilometer (billion ton-sea mile)	
	2008*	所占比例（%） Percentage	2008*	所占比例（%） Percentage
合 计 Total	**7 755**	**100**	**52 515**	**100**
原 油 Crude Oil	1 800	23.2	14 967	28.5
成品油 Refined Oil	575	7.4	3 206	6.1
铁矿石 Ironstone	858	11.1	8 361	15.9
煤 炭 Coal	830	10.7	6 284	12.0
谷 物 Corn	344	4.4	3 153	6.0
其 他 Others	3 348	43.2	16 544	31.5

注：*为估计数。
资料来源:《海运统计与市场评论》1月/2月 2009。
Note: *Estimated data.
Source: *Shipping Statistics and Market Review*, January/February2009, ISL.

11-17 主要沿海国家（地区）国际旅游人数
Number of International Tourists of Major Coastal Countries (Regions)

单位：万人 (10 000 persons)

国家和地区 Country and Region	国外游客到达人数 Arrivies Number of Foreign Tourists			出国旅游人数 Number of Tourists Going Abroad		
	2000	2006	2007	2000	2006	2007
世界总计 World Total	**68 763**	**85 289**	**91 147**	**76 040**	**103 579**	**110 037**
高收入国家 High Income	45 822	52 290	54 988	41 875	58 125	60 454
中等收入国家 Middle Income	20 866	30 246	33 023	24 700	32 609	36 531
低收入国家 Low Income	1 103	1 910	2 301			
中国* China	3 123	4 991	5 472	1 047	3 452	4 095
孟加拉国 Bangladesh	20	20	29	113	182	233
印 度 India	265	445	508	442	834	978
印度尼西亚 Indonesia	506	487	551		434	
伊朗 Iran	134	274		229		
以色列 Israel	242	183	207	353	371	415
日 本 Japan	476	733	835	1 782	1 754	1 730
哈萨克斯坦 Kazakhstan	147	347	388	125	369	454
韩 国 Republic of Korea	532	616	645	551	1 161	1 333
马来西亚 Malaysia	1 022	1 755	2 097	3 053		
缅 甸 Myanmar	21	26	25			
巴基斯坦 Pakistan	56	90	84			
菲律宾 Philippines	199	284	309	167	275	
新加坡 Singapore	606	759	796	444	553	602

11-17 续表 continued

国家和地区 Country and Region	国外游客到达人数 Arrivies Number of Foreign Tourists			出国旅游人数 Number of Tourists Going Abroad		
	2000	2006	2007	2000	2006	2007
斯里兰卡 Sri Lanka	40	56	49	52	76	86
泰 国 Thailand	958	1 382	1 446	191	338	402
土耳其 Turkey	959	1 892	2 225	528	828	894
埃 及 Egypt	512	865	1 061	296	453	
尼日利亚 Nigeria	81	111				
南非 South Africa	587	840	909	383	434	443
加拿大 Canada	1 963	1 827	1 793	1 918	2 273	2 516
墨西哥 Mexico	2 064	2 135	2 142	1 108	1 400	1 509
美 国 United States	5 124	5 098	5 599	6 133	6 366	6 405
阿根廷 Argentina	291	417	456	495	389	417
巴 西 Brazil	531	502	503	323	483	514
委内瑞拉 Venezuela	47	75	77	95	110	141
法 国 France	7 719	7 885	8 194	1 989	2 247	2 247
德 国 Germany	1 898	2 357	2 442	7 440	7 120	7 040
意大利 Italy	4 118	4 106	4 365	2 199	2 570	2 773
荷兰 Netherlands	1 000	1 074	1 101	1 390	1 670	1 756
波 兰 Poland	1 740	1 567	1 498	5 668	4 470	4 756
俄罗斯联邦 Russian Fed.	2 117	2 249	2 291	1 837	2 911	3 429
西班牙 Spain	4 640	5 819	5 897	410	1 068	1 128
乌克兰 Ukraine	643	1 894	3 212	1 342	1 688	
英 国 United Kingdom	2 321	3 065	3 087	5 684	6 954	6 945
澳大利亚 Australia	453	506		350	494	546
新西兰 New Zealand	178	239	243	128	186	198

注：*过夜旅客人数。

Note: *Data refer to overnight tourists.

11-18 集装箱吞吐量居世界前20位的港口
World Top Twenty Seaports in Terms to the Number of Containers Handled

单位：万标准箱 (10 000 TEU)

港　口 Seaport	所属国家或地区 Country or Region	吞吐量 Containers Handled
新加坡 Singapore	新加坡 Singapore	2 587
上海 Shanghai	中国 China	2 500
香港 Hong Kong	中国 China	2 104
深圳 Shenzhen	中国 China	1 825
釜山 Pusan	韩国 Republic of Korea	1 195
广州 Guangzhou	中国 China	1 120
迪拜 Dubayy	阿联酋 United Arab Em	1 112
宁波-舟山 Ningbo and Zhoushan	中国 China	1 050
青岛 Qingdao	中国 China	1 026
鹿特丹 Rotterdam	荷兰 Netherlands	980
天津 Tianjin	中国 China	870
高雄 Gaoxiong	中国台湾 Taiwan, China	858
安特卫普 Antwerp	比利时 Belgium	731
巴生港 Kelang	马来西亚 Malaysia	730
汉堡 Hamburg	德国 Germany	701
洛杉矶 Los Angeles	美国 United States	675
丹戎帕拉帕斯 Tanjung Periuk	马来西亚 Malaysia	600
长滩 Long Beach	美国 United States	506
厦门 Xiamen	中国 China	468
林查班 Laem Chabang	泰国 Thailand	462

资料来源：上海航运交易所，CL-ONLINE。

Source: Shanghai Shipping Exchange, CL-ONLINE.

11-19 货物吞吐量居世界前20位的港口（2008年）
World Top Twenty Seaports in Terms of the Cargo Handled, 2008

单位：百万吨 (million tons)

港口 Seaport	所属国家或地区 Country or Region	吞吐量 Cargo Handled	备注 Remark
上海 Shanghai	中国 China	581.7	内外贸货物 domestic and foreign trade cargo
新加坡 Singapore	新加坡 Singapore	515.5	内外贸货物 domestic and foreign trade cargo
鹿特丹 Rotterdam	荷兰 Netherlands	421.1	外贸货物 foreign trade cargo
宁波 Ningbo	中国 China	361.9	内外贸货物 domestic and foreign trade cargo
天津 Tianjin	中国 China	355.9	内外贸货物 domestic and foreign trade cargo
广州 Guangzhou	中国 China	347.0	内外贸货物 domestic and foreign trade cargo
青岛 Qingdao	中国 China	300.3	内外贸货物 domestic and foreign trade cargo
香港 Hong Kong	中国 China	259.4	外贸货物 foreign trade cargo
秦皇岛 Qinhuangdao	中国 China	252.3	内外贸货物 domestic and foreign trade cargo
大连 Dalian	中国 China	245.9	内外贸货物 domestic and foreign trade cargo
釜山 Pusan	韩国 Republic of Korea	241.7	内外贸货物 domestic and foreign trade cargo
南路易斯安娜 Southern Louisiana	美国 United States	233.7	内外贸货物 domestic and foreign trade cargo
名古屋 Nagoya	日本 Japan	218.1	内外贸货物 domestic and foreign trade cargo
深圳 Shenzhen	中国 China	211.3	内外贸货物 domestic and foreign trade cargo
光阳 Gwang	韩国 Republic of Korea	211.3	内外贸货物 domestic and foreign trade cargo
休斯敦 Houston	美国 United States	204.6	外贸货物 foreign trade cargo
安特卫普 Antwerp	比利时 Belgium	189.4	内外贸货物 domestic and foreign trade cargo
洛杉矶 Los Angeles	美国 United States	170.0	内外贸货物 domestic and foreign trade cargo
千叶 Chiba	日本 Japan	165.1	内外贸货物 domestic and foreign trade cargo
长滩 Long Beach	美国 United States	156.6	内外贸货物 domestic and foreign trade cargo

资料来源：Shipping Statistics and Market Review, December 2009, ISL.
Source: Shipping Statistics and Market Review, December 2009, ISL.

11-20 海上商船拥有量居世界前20位的国家或地区（2008年）
World Top Twenty Countries or Regions in Terms of the Number of Maritime Merchant Ships owned, 2008

国家和地区 Country and Region	艘数（艘） Number of Vessels (unit)	载重吨 Deadweight ton	
		万吨 10 000 tons	占世界% Percentage in the World Total
世界总计 World Total	**46 155**	**115 284.8**	**100.0**
巴拿马 Panama	6 842	27 087.9	23.5
利比里亚 Liberia	2 203	12 273.1	10.6
马绍尔群岛 Marshall Islamds	1 125	6 764.4	5.9
中国香港 Hong Kong, China	1 277	6 483.8	5.6
希腊 Greece	1 127	6 372.3	5.5
新加坡 Singapore	1 417	5 913.1	5.1
巴哈马 Bahamas	1 240	5 786.2	5.0
马耳他 Malta	1 487	5 076.0	4.4
中国 China	2 495	3 888.9	3.4
塞浦路斯 Cyprus	867	3 140.8	2.7
英国 United Kingdom	891	2 910.1	2.5
挪威 Norway	959	2 224.8	1.9
韩国 Republic of Korea	1 128	2 208.6	1.9
德国 Germany	522	1 803.5	1.6
印度 India	625	1 457.1	1.3
日本 Japan	2 524	1 446.2	1.3
意大利 Italy	779	1 422.0	1.2
安提瓜和巴布亚 Antigua and Papua	1 166	1 257.6	1.1
丹麦 Denmark	399	1 201.8	1.0
百慕大 Bermuda	142	956.7	0.8

注：1. 表中数据按载重吨排序。
2. 统计范围为300总吨及以上船舶，截止日期均为2009年1月1日。
3. 因统计口径不同，表中数据与其他出版物公布的数据稍有差别。

Note: 1. The data in the table are arranged in the order of deadweight tons.
2. The ships, ranging over 300 tons and more in gross ton, are counted as of Jan. 1, 2009.
3. Owing to different statistical requirements, the data given in the table may be somewhat different from those issued in other publications.

11-21 集装箱船拥有量居世界前20位的国家或地区（2008年）
World Top Twenty Countries or Regions in Terms of the number of Container Ships Owned, 2008

国家和地区 Country and Region	艘数（艘） Number of Vessels (unit)	集装箱位 Container	
		万标准箱 10 000 TEU	占世界 % Percentage in the World
合计 Total	**4 142**	**1 152.7**	**95.0**
巴拿马 Panama	798	257.7	21.2
利比里亚 Liberia	741	227.4	18.7
德国 Germany	330	121.2	10.0
中国香港 Hong Kong, China	222	69.0	5.7
新加坡 Singapore	284	67.3	5.5
英国 United Kingdom	203	67.2	5.5
安提瓜和巴布亚 Antigua and Papua	396	54.1	4.5
丹麦 Denmark	85	45.9	3.8
马绍尔群岛 Marshall Islamds	192	39.1	3.2
塞浦路斯 Cyprus	192	37.1	3.1
中国大陆 China mainland	174	32.2	2.7
美国 United States	82	25.2	2.1
希腊 Greece	45	21.6	1.8
马耳他 Malta	78	16.1	1.3
巴哈马 Bahamas	62	16.1	1.3
荷兰 Netherlands	83	16.0	1.3
法国 France	26	14.5	1.2
韩国 Republic of Korea	81	11.2	0.9
意大利 Italy	23	7.9	0.7
马来西亚 Malaysia	45	5.9	0.5

注：1. 表中数据按载重吨排序。
2. 范围为300总吨及以上船舶。
3. 因统计口径不同，表中数据与其他出版物公布的数据稍有差别。

Note: 1. The data in the table are arranged in the order of deadweight tons.
2. The ships, ranging over 300 tons and more in gross ton.
3. Owing to different statistical requirements, the data given in the table may be somewhat different from those issued in other publications.